THE GRANDFATHER CLOCK MAINTENANCE MANUAL

THE GRANDFATHER CLOCK MAINTENANCE MANUAL

John Vernon

VNR VAN NOSTRAND REINHOLD COMPANY
New York Cincinnati Toronto London Melbourne

Drawings by A. Blackbourn
Photographs by W. Payne

Library of Congress Catalog Card Number 83-10502
ISBN 0-442-28827-1

Printed in Great Britain

Published by Van Nostrand Reinhold Company Inc.
135 West 50th Street
New York, New York 10020

16 15 14 13 12 11 10 9 8 7 6 5 4 3 2 1

Library of Congress Cataloging in Publication Data

Vernon, John, 1904–
The Grandfather Clock Maintenance Manual.

Bibliography: p.
Includes index.
1. Longcase clocks—Maintenance and repair. I. Title.
TS547.V47 1983 681.1'13 83–10502
ISBN 0-442-28827-1

CONTENTS

INTRODUCTION

This book aims to help aspiring repairers to look after a grandfather clock. Living grandfathers and longcase clocks alike are generally treated with care and respect – the latter perhaps more than the former. The object here is to include the information necessary for anyone undertaking the care, maintenance and even restoration of the commoner forms of 8-day longcase clock, with a section also on the 30-hour clock. (The more complicated chime and musical clocks are for specialists only.)

The work described is suited to people who are not specially qualified but who know their way round a workshop. Any student of a craft has to learn a few of the technicalities and to acquire a few tools; the serious student who does not have a lathe will readily find a horologist or model-maker who will help him out. For people unfamiliar with the necessary horological terms, a simple glossary is included. The amateur should also, of course, study various types of clock, wherever and whenever possible, and examine in detail the pictures in clock books.

The qualified horologist may discover nothing new in this book, but if he likes clocks he may get pleasure from it, even if only in reading – and perhaps being able to criticise – someone else's views on his pet subject.

If the book sets someone on the road to being a craftsman, so much the better. Take your time and enjoy the work – but as an old craftsman once said to me, 'It's no use setting about mending clocks unless you're in the right mood.'

THE ENGLISH 8-DAY LONGCASE CLOCK

The main parts of a longcase clock are the movement, the weights and pendulum, the trunk of the case and the hood.

The movement is mounted in the case on a board, known as the seat-board, and this carries the whole weight of the clock including the pendulum and driving weights, in all some 40 to 45lb (about 18 to 20kg). On some clocks the seat-board is fastened down, which is not strictly necessary. The seat-board is generally supported by two vertical boards at right angles to the back-board of the case. A weight of 40lb (18kg) or so will 'sit' firmly on these 'cheeks' without moving in any way.

The pendulum, which really tells the time, is suspended from a brass piece at the top of the back of the movement. The clock merely counts the swings and restores the momentum lost by the pendulum as it oscillates.

The longcase clock depends entirely on gravity for its operation. Gravity is at any one place a constant force, so it should be reasonably easy to apply such constancy to a machine capable of keeping accurate time. This study has interested man since he discovered the lever, but it does not immediately concern us here.

Let us therefore briefly examine the parts of the clock, taking them in the order mentioned above.

The Movement

For our purpose, reference is to the grandfather clock only, but much of what follows is applicable to most mechanical clocks.

The three main sections of the movement are the going train, the strike train and the motion work. The trains are mounted between two brass plates, and the motion work is between the front plate and the dial plate.

a) *The going train* consists of a weight, a line and a pulley, the barrel, which is fixed to the great wheel carrying a pawl called a 'click', which works on a ratchet cut round the rim of the end-plate of the barrel, the centre wheel, the third wheel, the escape wheel, and finally the pallets arbor and crutch. The latter imparts the impulse to the rod of the pendulum, and the train moves all the time that the clock is going.

b) *The strike train* which moves only when the clock strikes, or when manually set off, consists as before of a weight, a line and pulley, a barrel and great wheel with its click assembly. Then comes the pin-wheel having eight steel pins mounted at right angles to, and near, the rim and equidistant from each other. They operate the bell-hammer. Next comes the pallet wheel, which has an arbor extended through the front plate, and on its squared end is fixed a one-toothed pinion called the gathering pallet. Then comes the warning wheel with one pin protruding from its side; and finally the fly, which is a simple form of air-brake to prevent the train running too fast. The blades of the fly are friction-fitted on to the arbor.

c) *The motion work* consists mainly of three wheels, but to describe only these might give the wrong impression. It is advisable, therefore, to describe all the parts fixed to the front plate of the clock behind the dial. The centre wheel arbor of the going train is the longest in the clock, and it protrudes through the front plate for some 2in (50mm). Then comes a tension washer, the minute wheel and the bridge, on which is mounted a short pipe

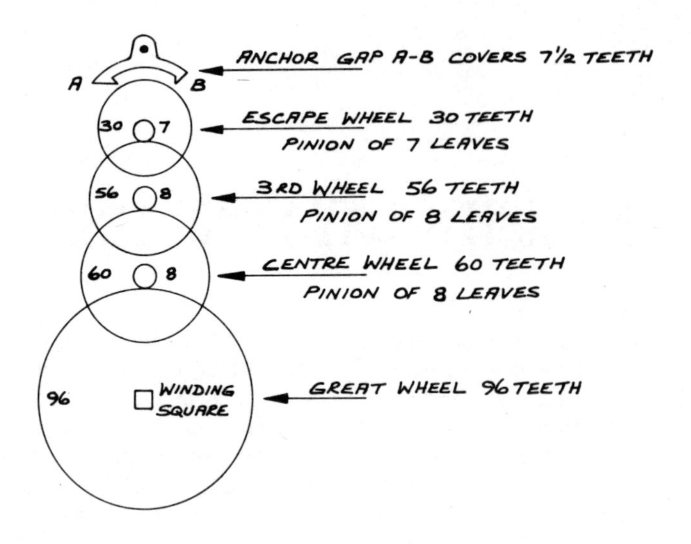

Figure 1

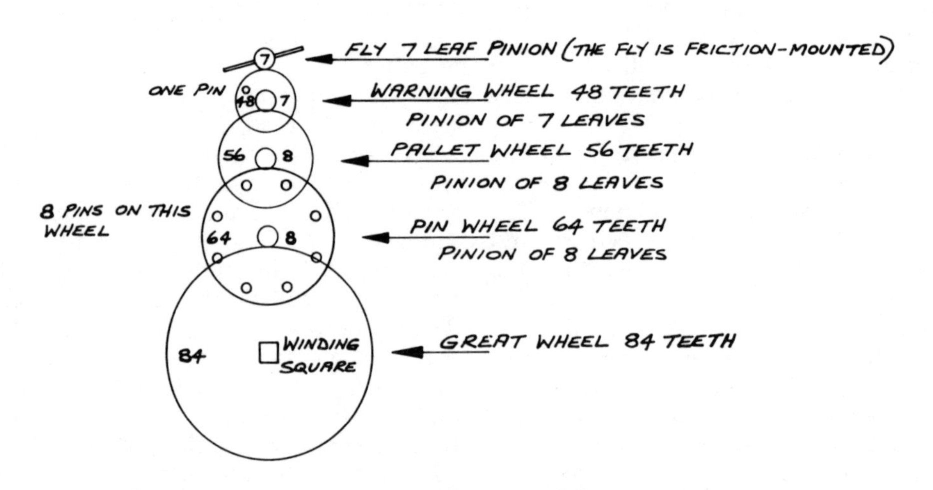

Figure 2

which acts as an axle to the hour wheel, all of which will be described later in more detail.

On the motion wheel, extending to the front, is a short steel pin which operates a V- or L-shaped lever called the lifting-piece. This is mounted at its angle-end on a post or stud screwed into the top right-hand corner of the front plate in such a manner that it acts like a bell crank. There is also another lever called the rack hook, mounted on the left side of the front plate opposite the lifting-piece. The hook of this rests in a toothed ratchet called the rack, which is a third lever pivoted on a post mounted a little to the right of the strike-train winding square.

At this stage the student needs only to recognise these parts and to get to know their names. The function of each will be described later on as the need arises. So much, therefore, for the parts of the movement.

Date wheels, moon-work, etc, are covered later.

The Trunk

This is the main body of the clock case. It is in three sections, the top, the body and the plinth. The top is constructed to hold the hood, which slides off to the front. The body has a door in its front, and sometimes a bolt or wooden 'spoon' to latch the hood in place. The width of the body should be roughly the same as the width of the clock dial. The hood and plinth are wider, much the same as each other, proportional balance permitting. The plinth at the base is the part on which the whole clock stands; and nearly all the plinths of clocks more than a hundred years old have been mended, added to, or more often cut down to shorten the clock to fit into modern houses or flats.

The Hood

One could almost describe this as a three-sided box with a glass front which covers the clock movement. It and the trunk are often decorated, sometimes ornately, less often grandly and generally sedately.

It may have brass finials at the top—ball and spire at the sides and ball and eagle at the centre, or a larger central spire. Sometimes these are of gilded wood. Over the top of the hood door there may be a silk-lined fret, cut to let out the sound of the bell more readily.

At the hood sides there are two or four plain, fluted or twisted pillars, each being set in a brass capital or end-piece. The top of the hood may have two opposed swan-necked pieces, and on the front of each may be mounted a brass ornament in the form of a rosette, or sometimes a figure. For instance, a clock made in 1805 or 1806 could well portray the head of Admiral Nelson.

The trunk and plinth of the case are backed by a long board, usually made of elm wood, which extends upwards to form the back of the hood also, when the latter is in place. There should be a close fit between hood and back-board to prevent the entry of dust.

The foregoing is a brief but temporarily adequate description of the parts of a longcase or grandfather clock. As you study these clocks, you may like to re-read it occasionally, and to consult the glossary of terms at the end of the book.

DISMANTLING, CLEANING AND REPAIRING

Let us assume, now, that you are about to tackle a grandfather clock which needs attention.

1 First of all, study the clock carefully and note down anything out of the ordinary. Open the trunk door observing the type of fastening, the lock, key, etc, and whether it needs attention.

Record the name of the maker of the clock for research and to help in dating the clock. Pocket the winding key, and remove the weights. If these are low down within the case, wind them up so that you may readily get at them. Do *not* wind them up fully, because if you do you will find it difficult to unloose them. Note that the lines are wound properly on the barrels: if they are dislodged and become wound round the arbors it is a ticklish business to free them. This happens sometimes if the clock – generally one with a shortened plinth – has been allowed to run right down. The weight, due to the shortness of fall, has touched the floor and thus slackened the line. If the clock is then wound up the line may readily get wound round the arbor.

2 Next, examine the hood, which may be fixed inside by a bolt or wooden catch. There are several varieties of this catch, some holding the door of the hood closed as well as holding the hood itself on to the trunk body. Release the catch and remove the hood by sliding it forward towards you. Some very old clocks have hoods which lift up vertically but they are now very rare.

Watch the sides of the hood, especially if it has glass windows, and take care not to let the hood door swing out – it is easily broken if it hits something. Hold it firmly when removing it because it could tilt off balance. Remove the dust and place the hood carefully to one side out of the way. If the glass needs replacing, see the Appendix. On a decorated clock, check whether the finials are loose, and note the state of any veneers or marquetry.

3 Now remove the pendulum by gently lifting the rod with one hand while the other frees the suspension spring at the top. Take care not to lift the rod too much or to damage either the spring or the wire crutch through which the rod passes. If the clock is standing obliquely across the corner of the room, and it is not desired to move it, you may require steps to see what you are doing. Note that the pendulum is made up of several parts, the average being six, as follows: a, the suspension spring; b, the brass block; c, the rod, of steel wire, wood or brass or iron flat strip; d, another brass or iron block, squared or tapered; e, the pendulum bob; f, the rating nut, which works on a threaded short rod let into the base of the squared end-piece.

4 If the seat-board is fastened down, remove the nails or screws taking care not to damage the wood in any way. Lift the movement clear of the case, making sure that the lines and pulleys do not damage the wood or veneers on the way out. The case of the clock is now empty, and presumably standing in its original position. Test it for firmness of standing. Check whether the 'cheeks' or supporting boards on which the seat-board rests are level. If they are not, note just how

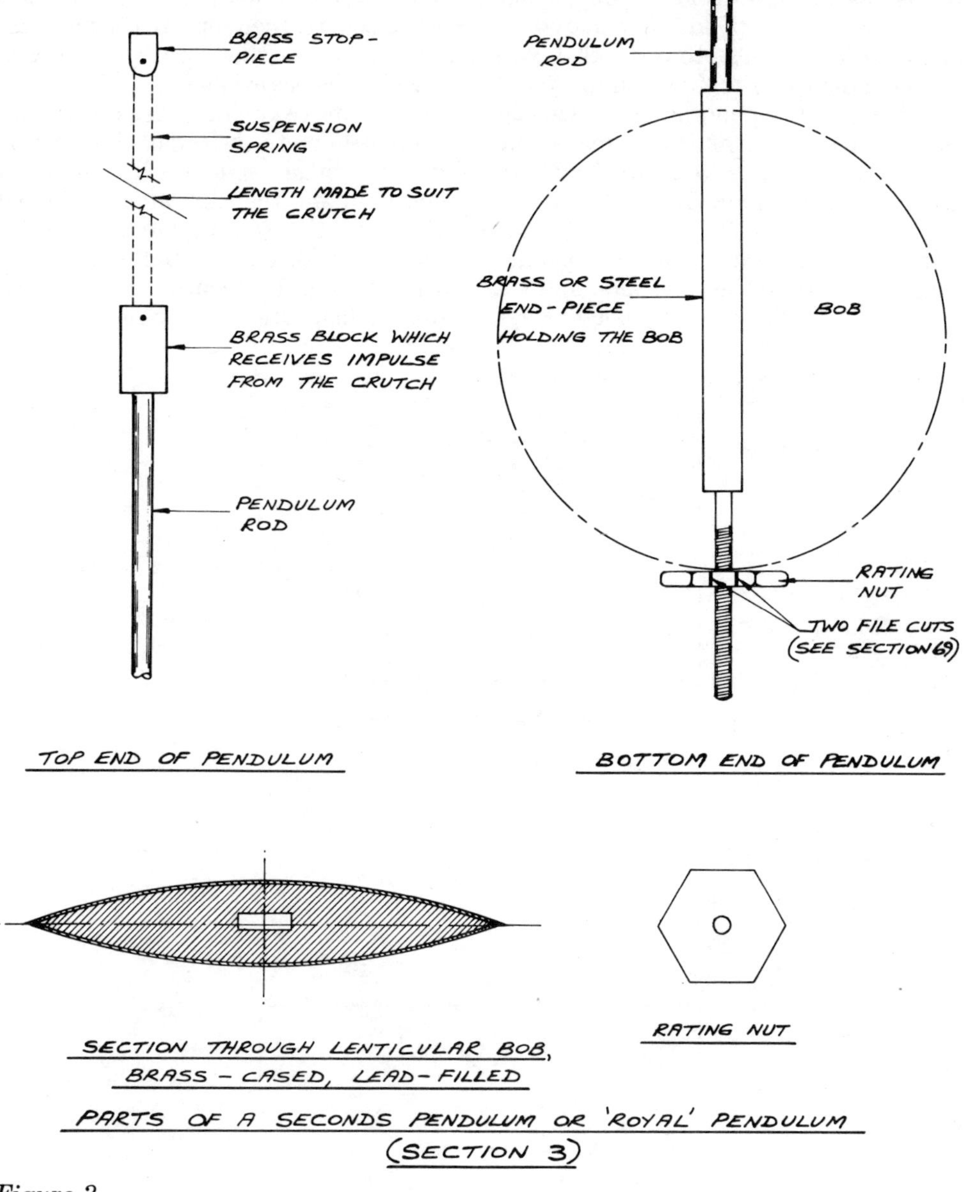

Figure 3

unlevel they are, and the direction of the tilt, because this may be useful later on. The condition of the cheeks, or the discovery that they are badly out of true, may mean that repairs will be necessary before re-setting the seat-board. If not, replace the hood and secure it in place. If the clock is not yours, put a label with the owner's name on key and movement.

5 Before proceeding further, provide yourself with a few simple requisites – cleaning rags, a stiff cardboard box strong enough to support the movement and approximately 5 × 4 × 3in (13 × 10 × 8cm) in size, various trays on which to lay out the parts, a 6in (15cm) diameter bowl half-filled with kerosene, an old toothbrush, some fine steel wool, and the ordinary everyday tools that the handyman normally uses.

As to special tools, a small lathe or

turns is essential, so that if you do not possess one — they are expensive — arrange to have access to one. Also, you will need a depthing tool, which may cost from £65 to £125 depending upon size and accuracy. You may, however, make one for yourself, but here you must refer to the textbooks (see page 96).

6 Now take the movement, key, pendulum and weights to the workshop and set the former upright on the bench. First remove the bell, by slackening the nut on its top centre; then replace the nut loosely and remove the steel bell-bracket, replacing its screw also.

Now remove the back-cock — a kind of bridge held on to the top of the back plate by two steel screws; it carries the pendulum and the rear pivot of the pallet arbor. As the screws are loosened and removed, take care to pull off the back-cock straight backwards — two steady-pins prevent lateral movement.

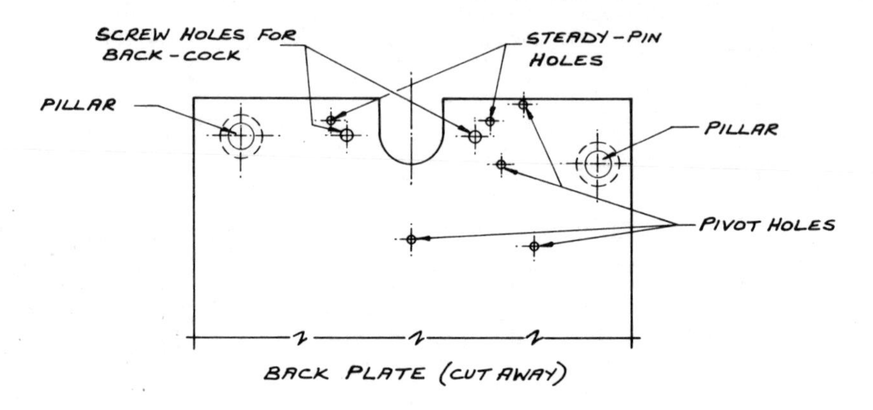

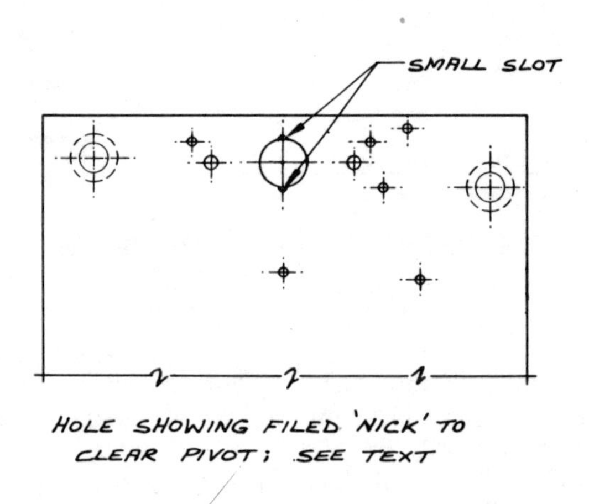

BACK PLATE OF MOVEMENT (SECTION 6)

Figure 4

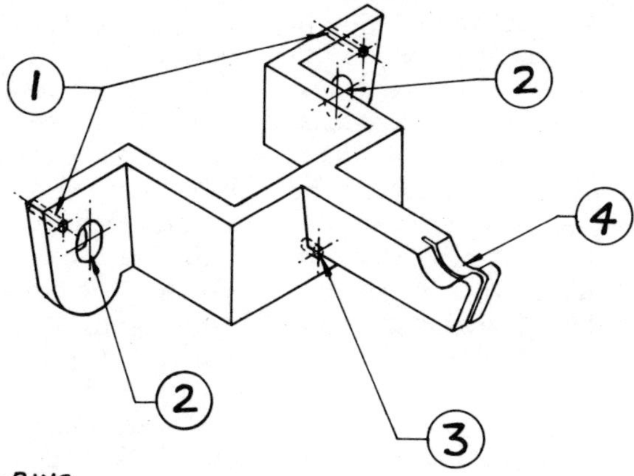

1 STEADY-PINS
2 SCREW HOLES
3 REAR PIVOT HOLE FOR PALLET ARBOR
4 PENDULUM SUSPENSION

NOTE: THE PIVOT HOLE 3 IS ON A LEVEL WITH THE BEND IN THE SUSPENSION SPRING, OR NEAR THERETO. THIS REDUCES TO A MINIMUM ANY SLIDING MOTION BETWEEN THE FORK OF THE CRUTCH AND THE BRASS BLOCK AT THE TOP OF THE PENDULUM ROD

BACK-COCK (SECTION 6)

Figure 5

Do not turn the back-cock at all until it is clear of the pivot of the arbor carrying the 'anchor' or pallets. The pallets may now be moved backwards to clear the front pivot hole. The pallets, complete with the crutch attached, should now be carefully removed by turning so as to allow the crutch to be lifted gently through the hole in the back plate. Sometimes if the hole is a small one it will be found that a 'nick' or small slot has been cut in its lower edge (sometimes upper edge) to permit clearance for the rear pivot as it passes through. Again, in some clocks instead of a hole there is a U-shaped opening in the top of the back plate; in that case, after freeing the front pivot, the pallet arbor can be lifted straight out. Place the parts in the kerosene bowl (see Figure 4).

Now lift the movement, turn it dial-upwards and rest it on the small cardboard box, with the underside of the seat-board towards you. Push the ends of the cables or lines so that the knots are clear of the board top, and either remove or undo these knots. Pull the lines through the board and remove the pulleys, placing both safely in the bell, and the bell on one of the trays.

Seat Bolts

7 Observe how the movement is secured to the seat-board. The two most frequent methods are either for two bolts to be screwed directly from beneath the board into the two lower columns, or to have two hooked bolts, which instead of screwing into the columns hold the latter down by their hooked ends. Each hook is secured beneath the board by a wing-nut or similar form. In the early days there was no means of tapping a thread in a hole as deep as that in a column centre; so a hole of one size was drilled halfway through the column and a slightly smaller hole was drilled through the rest of the column. The smaller hole was tapped to take the bolt and it is generally found

that in a column the threaded portion is the uppermost. The nuts or bolts, as the case may be, beneath the seat-board should be slackened off, and the hooks or bolts carefully removed. Hold the board in one hand while doing this and undo the nuts with the other, because the clock will be on its back and you will have to avoid any falling apart when you loosen the bolts. Pull the seat-board clear and free of the lines, and put the pulleys and bolts in the bell.

8 Before we leave the subject of seat bolts, another point is worth bearing in mind. Most of the columns or pillars on old clocks have an enlarged central portion turned to a disc-shape or sometimes a ball-shape. If the bottom pillars are near the edge of the plates, this central part may be found flattened on the underside so that the 'flat' is level with the bottom edges of the plates. Thus when the clock is placed on the seat-board, all three surfaces are in contact with the board. Any bolt screwed into such a column in the centre of the flat portion will, when tightened, exert no torque on the column, which is ideal.

If your clock has a clearance between the column and the seat-board, any *undue* pressure downwards on the column or pillar may produce torque. The ends of the pillar held in the plates resist any pull downwards by the bolt or hook, and the pillar transfers pressure to the plates. This torque tends to set up stress in the plates and could upset the delicate setting of pivots and pivot holes. The golden rule, therefore, is to tighten the holding-down bolts just enough to hold the movement firmly, so that the plates will not shift even under the constant vibration of the striking hammer of the clock, or during winding.

Where a bolt is screwed into a pillar liable to torque, the latter may sometimes be avoided by inserting a washer of suitable thickness between the column and the seat-board; but this should not be necessary and is not recommended. A reasonably wide washer positioned under the head of the holding-down bolt will protect the wood on the underside of the board. Where the column is held by a hook, care must be taken because these hooks are not central, pressing on one end of the pillar more than the other. If the hook for the left column is to the rear end, then the right-hand hook should be towards the front of the right column, so balancing the hold.

If the seat-board is wormed, warped or badly split, it should be replaced by another of fully seasoned good-quality timber, as thick as, and cut to the pattern of, the old one, so that it will fit back into the clock case without trouble. The clearance slot or hole for the pendulum rod is generally found to be unnecessarily big. A new board can be made to rectify this.

9 It is a good idea to have a notebook handy, for by now we are ready to examine the movement for anything broken or badly worn, and to note it down, especially broken wheel-teeth or pinion leaves, worn cables, clicks, pivot holes, etc; and this done, we can then set about removing the hands and the dial of the clock. Notice we refer to the dial, not the face, of a clock.

Removing the Hands

10 First gently pull off the seconds hand, which is a friction fit on the front end of the escape-wheel arbor. Steady the escape wheel with one hand while you do this, and remove the seconds hand with

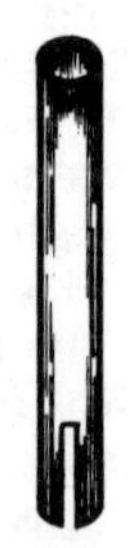

1/8 INCH PIPE SLOTTED BOTH SIDES

TOOL FOR REMOVING HANDS (SECTION 10)

Figure 6

Break-arch longcase clock by Henry Raworth, Plymouth. It has urn spandrels and attractive dolphins in the arch. The hands are too long to be original and the date ring has been added—note the blocked lower part of the aperture. The case is mahogany, with turned columns and a break-arch top to the hood door. The centre finial is broken (Section 1)

Seventeenth-century, 30-hour grandmother clock, 74in (188cm) high. It has a single hand, and alarm ring in a centre setting. The dial is marked with a quarter-hour ring (no minutes) and decorated with mask spandrels and centre engraving. The oak case has a hood door with integral pillars the other. With a small tool which you may readily make for yourself (see Figure 6), like a little tube with a slot cut in one end, press down the hand collet with the slotted end of the tool clearing the pin, and remove the latter with a small pair of pliers. Lift off the hand collet and also the minute hand, then finally the hour hand, which may be held by a small retaining screw.

For the smaller parts of the clock you will find a flat, light-coloured small tray or dish very useful; preferably one with curved or bevelled inside edges for ease in removing things. Such parts can be placed on the trays and put away safely while you work on other parts.

11 On many clocks, more often those with painted dials, there may be a date dial between the hands centre and the VI on the chapter. This dial has a date hand which is attached to a wheel mounted behind the dial plate, so there is actually no need to remove the date hand – it will remain attached to the dial when you remove that.

Sometimes there is an arched aperture instead of the date dial, the date being indicated by figures shown through the opening. The wheel bearing the figures is also mounted on the dial plate and need not be removed yet.

Hand Mounting

12 The hour hand is sometimes a friction fit on the pipe of the hour wheel, the hand being mounted on a collar of brass or copper made to fit the pipe. On a clock where a friction washer has been fitted behind the hour wheel, the hour hand is squared on to the pipe of the hour wheel; if this latter fitting is at all loose, it may be repaired by drilling very fine holes across two opposite corners of the square, so that very thin clock pins can be inserted over the boss of the hour hand to hold it firmly in place. Again, if any three corners are slotted with a fine saw, a thin horseshoe-shaped washer may be slid in place to hold the hand.

A refinement of this method is mechanically sounder; and the washer is made with a square hole exactly to fit the square of the pipe. In this case four slots are cut, one in each corner of the square, and the washer is placed over the square, pressed down on the hand-boss and then rotated to engage the slots (see Figure 7).

In good-quality clocks the large friction washer mounted behind the hour wheel enables the hour hand to be moved in relation to the wheel (but not to the snail) and hence to the minute hand. The hour hand can thus easily be set in its correct position. Where there is no such washer, any attempt to turn the hour hand while the minute hand is held still may well result in breaking the hour-wheel teeth. So where this is the case, take care when reassembling the clock to set the hour wheel in its correct position vis-à-vis the pinion on the motion wheel (see Section 50).

Dial Removal

13 With the hands of the clock removed the dial is ready to be taken off, and you will see that it is held to the front plate of the clock by four (sometimes three) pillars or columns which are pinned on the inside of the front plate. Removal of the pins will allow the dial, complete, to be withdrawn. If the pins are rusted-in or corroded apply some penetrating oil and leave for a few minutes before trying to extract them. A long, thin, concave-ended steel punch will be found useful here.

Most brass dials, having separate chapter rings, date rings, spandrels, etc, are mounted direct on to the front plate of the clock, as described above. Most painted dials, however, have a 'false' plate (see Section 17) between the front plate of the clock and the dial proper. This false plate or false dial is made of cast iron and may bear the foundry name. It has four holes to fit the dial pillars, and three or four columns of its own to fit the clock. Where such false dials are present the whole dial and false dial plate should be removed together, especially if there is a break-arch dial with moon-work attached to it (see Figure 8).

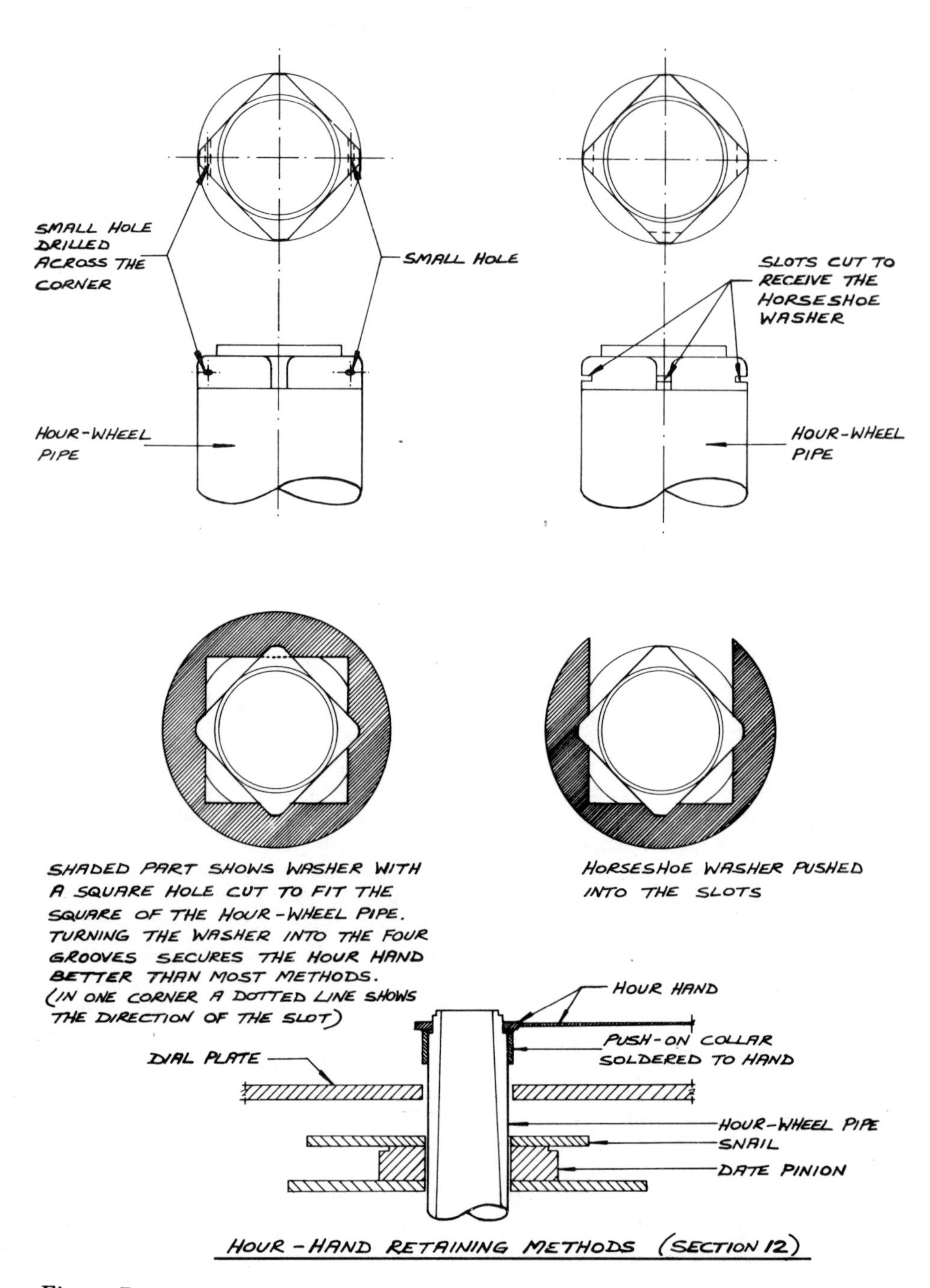
SMALL HOLE DRILLED ACROSS THE CORNER
SMALL HOLE
SLOTS CUT TO RECEIVE THE HORSESHOE WASHER
HOUR-WHEEL PIPE
HOUR-WHEEL PIPE
SHADED PART SHOWS WASHER WITH A SQUARE HOLE CUT TO FIT THE SQUARE OF THE HOUR-WHEEL PIPE. TURNING THE WASHER INTO THE FOUR GROOVES SECURES THE HOUR HAND BETTER THAN MOST METHODS. (IN ONE CORNER A DOTTED LINE SHOWS THE DIRECTION OF THE SLOT)
HORSESHOE WASHER PUSHED INTO THE SLOTS
HOUR HAND
PUSH-ON COLLAR SOLDERED TO HAND
DIAL PLATE
HOUR-WHEEL PIPE
SNAIL
DATE PINION
HOUR-HAND RETAINING METHODS (SECTION 12)

Figure 7

Types of Dial

14 Set the actual movement aside for the moment, and inspect the dial assembly for missing or broken parts, damage or blemish. There are several types of dial and the variety of each is legion, so that the serious student should study detailed books.

The dials of longcase clocks fall mainly into the following groups:

a) Ten-inch square brass dials for 30-hour clocks having four spandrels, a chapter ring and sometimes a date ring, and an aperture above the VI. There are no winding holes or seconds hands. Some are 11in square.

b) Twelve-inch square brass dials for 8-day clocks, with chapter ring, date ring, four spandrels, a matted centre with two winding holes, and of course a date aperture.

c) Twelve-inch break-arch dials, brass with six spandrels and a roundel in the break-arch. The roundel is sometimes a calendar dial or strike-silent dial; or perhaps just decorative – engraved 'Tempus fugit'. Some of these depict the figure of Chronos – Father Time – armed with a scythe and heavily hirsute.

d) As (c) above but with moon-work, ie a large wheel painted with two moon faces, sometimes with silvered edges, and the date of the moon's age, revolving behind two 'hemispheres' engraved as the world – all in the break-arch. The hemispheres are set side by side with the pivot of the moon wheel between them; and they are integral with the dial plate. Either the rim of the moon wheel, or the arch frame just above it, is marked off to show 29½ days, and the wheel will be found to have double that number of teeth to make a whole number, and then again doubled to allow for movement twice per day, and depiction within a half-circle only. Thus there will be 118 teeth in the moon wheel, and two 'faces' – it revolves once in two months.

e) Twelve-inch square painted dials with seconds hand and either a date hand or a date wheel showing through an aperture beneath the centre hole.

f) Twelve-inch break-arch painted dials with decorative pictures in the corners and break-arch; the corners often portray the four seasons.

g) As (f) above but with painted moon-work with very much the same layout as for brass dials. There are break-arch dials which have automata in the arch, showing for instance a ship rising and falling with the tide, or more commonly rocking on the waves. The times of the tides at various ports may be indicated: eg a Bristol or Plymouth made clock would show the times of high tide at those places.

Rarer are the ten-inch painted dials for 30-hour clocks, usually of an early nineteenth-century type. Yet another type of dial is of engraved brass, silvered all over, for which only one plate is involved, the sole decoration being the engraving and the overall silvering.

15 Painted dials cannot be dismantled except as already described, but brass dials can; and the details are common to most.

First extract the holding pins at the back of the dial plate and remove the chapter ring, seconds ring, name plate, and boss or roundel if any. Remove the spandrels and replace the screws in each for safety.

Now remove the retaining wheels of the date ring and the date ring itself. Some date rings revolve in brass cocks screwed behind the dial plate, and one of these usually turns aside to allow the date ring to be taken out. This type is found mainly on brass dials, seldom on painted dials, and on the latter has sometimes been added.

Moon-Work

16 If the break-arch of the dial plate is mounted with moon-work, called lunar work, first remove any levers or 'jumpers' which operate the moon wheel, and finally the moon wheel itself. The latter is mounted on a stud – a miniature stub-axle – fixed to the dial plate between the two 'hemisphere' discs mentioned earlier. This stud is very often found to be bent, from previous rough handling. It should ideally be at right angles to the plate, and in re-setting it care must be taken to see that it is firm, that the wheel revolves without wobble or friction, and that proper all-round contact is made with the jumpers and levers.

Where a lunar wheel has been bent it must be straightened very carefully – often a difficult task. If you damage the paintwork it is costly to get it renovated. Care should also be taken not to damage the paint on the main dial where the stud is fixed, though this can readily be touched up by a reasonably skilful person.

False Dials

17 When replacing false dials on painted-dial clocks, see that all the joints are firm and steady; the dial plate should not move when in position. It must be remembered that the bell-hammer shakes the movement every hour, day and night, and that the vibration can easily loosen and shake out a badly fitting pin. Where the false dial has been cast with the holes for the columns already provided, the holes are very seldom a close fit – being cast and not drilled. They are therefore larger than the dial 'feet' which they

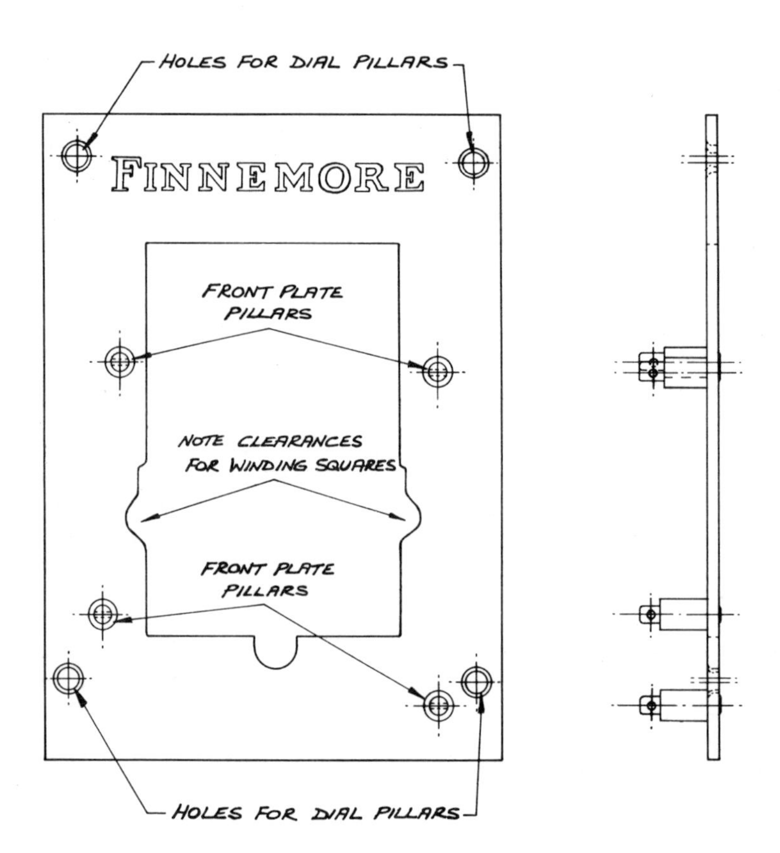

FALSE DIAL SHOWING MAKER'S NAME

(SECTIONS 13 & 17)

Figure 8

accommodate, so obviously the holding pins should be long enough to ensure a firm hold.

Dial Cleaning

18 The various parts of the dial are now ready for cleaning and there are many ways of doing this. Always remember that *any form of abrasive must be avoided.* Various special cleaning fluids may be obtained from any supplier of horological materials. One of these is most suitable for non-ferrous parts, and has to be diluted — one part solution to seven parts of clean water. Many of the brass parts, including the spandrels, will have been lacquered, or gilded and lacquered, and the lacquer will by now have perished, worn thin, oxidised or become unsightly. Soaking the pieces in the solution, for the time specified by the manufacturer, will remove all old lacquer and tarnish, and will bring up the full beauty of the gilding. On brass however it leaves a matt finish which is easily shined with a soft cleaner and finished off with a jeweller's brush. The parts being treated must be totally immersed in the solution, no part breaking the surface. Where brass wheels are mounted on ferrous arbors, the latter will come to no harm; in fact they are often improved. *Do not put any of the rings into the solution* because it may remove the black wax.

If any of the chapters are faulty, having already lost some of the black wax, this can be remedied, but it is a fairly tricky business — fully described in the textbooks (see Reference Sources, page 96). One method is to remove the old wax, clean the cavity and refill it with melted resin to which has been added ordinary lampblack. This makes a black sealing-wax and its use will over-fill the engraving of the chapter being worked on. If the whole chapter ring is warmed in a low oven the wax will soften to fill the hollow.

When cooled, the ring can be gently 'grained' with a very fine wet pumice block, and the surplus wax ground away to leave a sharp edge and a clear figure. Further warming, just enough to melt the wax to a shiny finish, will complete the job. The student is especially advised first to try out this process on odd bits of engraved brass if he has any, and to master it thoroughly before attempting to restore a clock dial.

During graining any silvering will have been destroyed, so that the ring will need to be re-silvered. Again this is a delicate process, well described in the textbooks. It should be practised to perfection before you attempt to re-silver a chapter ring or clock dial (see Reference Sources, page 96).

We have, of course, digressed into the realms of the expert, but the student has been fully warned; and in any case, let this glimpse into that realm tempt him gradually to explore further.

Examination of the Movement

19 Returning to the movement of the clock, examine it again for breakages or cracks, especially in the ends of the pillars; and see that no wheel-teeth are missing. All the pinions must be sound and not unduly worn. Check your notes to see that all defects are recorded — when there are some two or three dozen jobs to be done on a clock, each one should be ticked off a list as it is completed.

If the movement is black with grime and dirt, as so many of them are, all parts will need to be washed in the kerosene bowl, brushed clean and dried before being placed in the cleaning solution referred to earlier. On removal from the solution wash the parts well in hot water; the heat will hasten the drying time, which obviates oxidisation or rust.

The examination of parts is much easier if you have an eyeglass lens of about 3½ to 4in (90 to 100mm) focal length.

Motion Work

20 The hour wheel and others which are mounted on the front plate of the movement, between it and the dial plate, are known as the motion work, as already mentioned. The arbor which carries the minute wheel is the longest in the clock, and it protrudes through the front plate of the clock for some 2in (50mm) or so. It passes through a closely fitting but freely rotatable little pipe, one end of which is squared to carry the minute hand; to the other is fixed a solid brass wheel, which is toothed to engage with another wheel of similar diameter and with the same number of teeth. The latter is mounted on a six-leaved brass pinion rotating on a stud.

There are two ways of naming these wheels. One is to call the wheel on the pipe the 'minute' wheel, and the one it contacts the 'motion' wheel; and the other way is to call the former the 'cannon' wheel – its mounting giving a vague suggestion of a miniature cannon, perhaps – and the latter the minute wheel. The first of these two pairs of names seems preferable, if only for logical reasons: the hour hand is mounted on the hour wheel, which has no other name, so why not call the wheel which carries the minute hand the minute wheel?

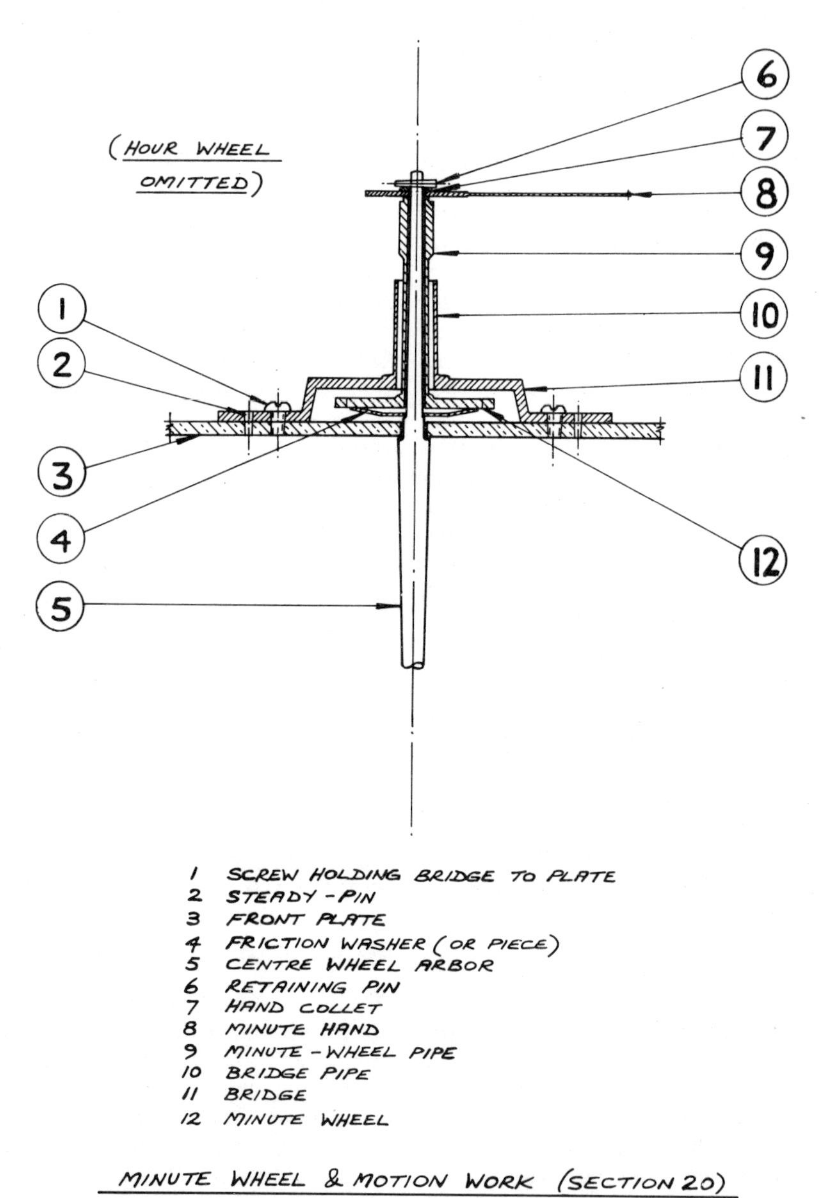

Figure 9

The motion wheel, also sometimes referred to as the intermediate wheel, is mounted on a steel stud fixed in the front plate so as to allow the wheel to engage with the minute wheel at the same time as its pinion engages with the hour wheel (which has 72 teeth) (see Figure 12).

The motion wheel, being the same size and having the same number of teeth as the minute wheel, must rotate at the same rate. Thus when the minute wheel turns 12 times the motion wheel does so too, and its six-leaved pinion engages all 72 teeth of the hour wheel, which of course has turned but once.

Now if the minute wheel was rigidly fixed to the centre arbor, it would not be possible to move the hands of the clock to re-set them; and to explain this further it will help to describe how this is overcome.

The centre arbor — the long one carrying the centre wheel — has a small

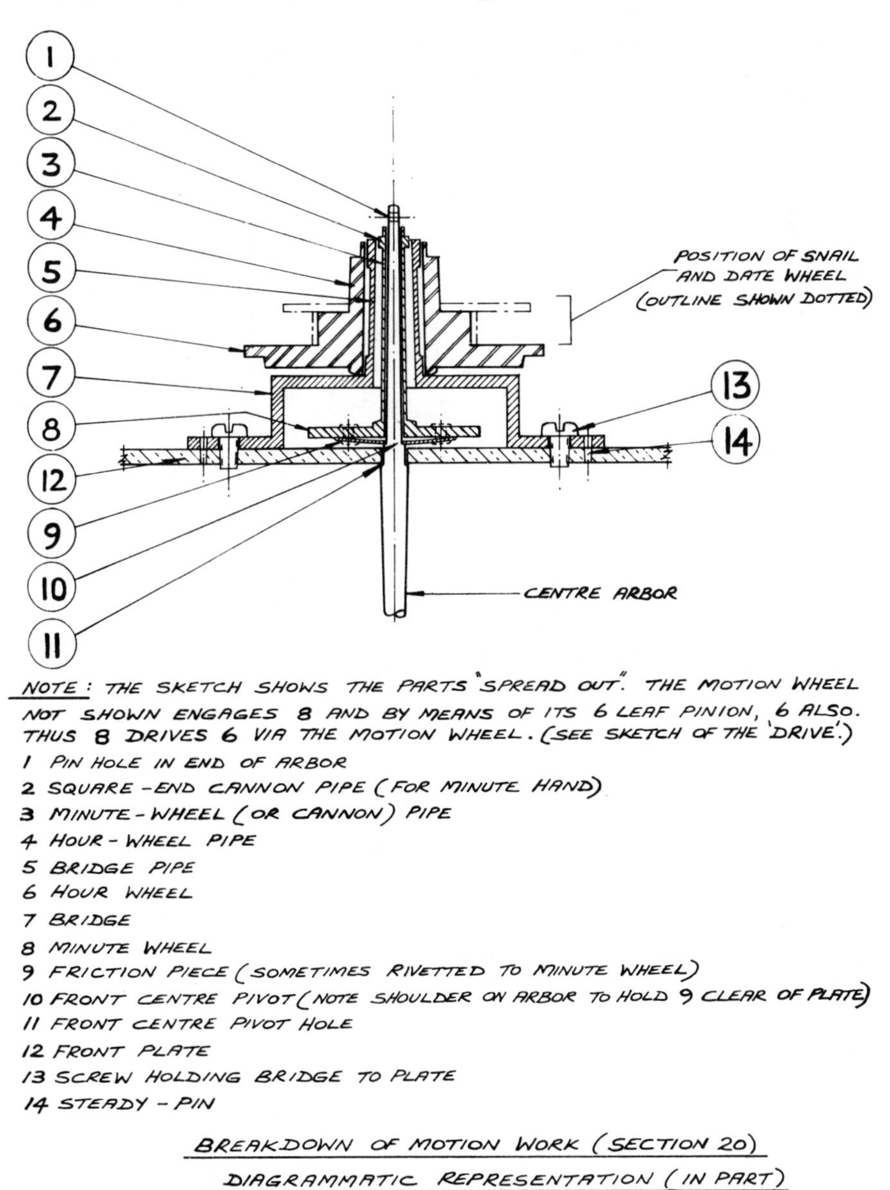

Figure 10

shoulder, generally squared, just in front of the front plate. An oval-shaped piece of hardened thin brass, springy and bowed a little, has a square hole in its centre. It is in fact an oval washer with a square hole in it, and it fits over the square on the arbor with its concave side next to the minute wheel. The shoulder forms a little ledge over which the washer cannot pass; so it is prevented from touching the front plate of the clock. Thus when the minute hand on the minute wheel is pressed down and pinned on, the washer is compressed and it transfers rotation to the minute wheel by friction.

When the clock is re-set to time, the

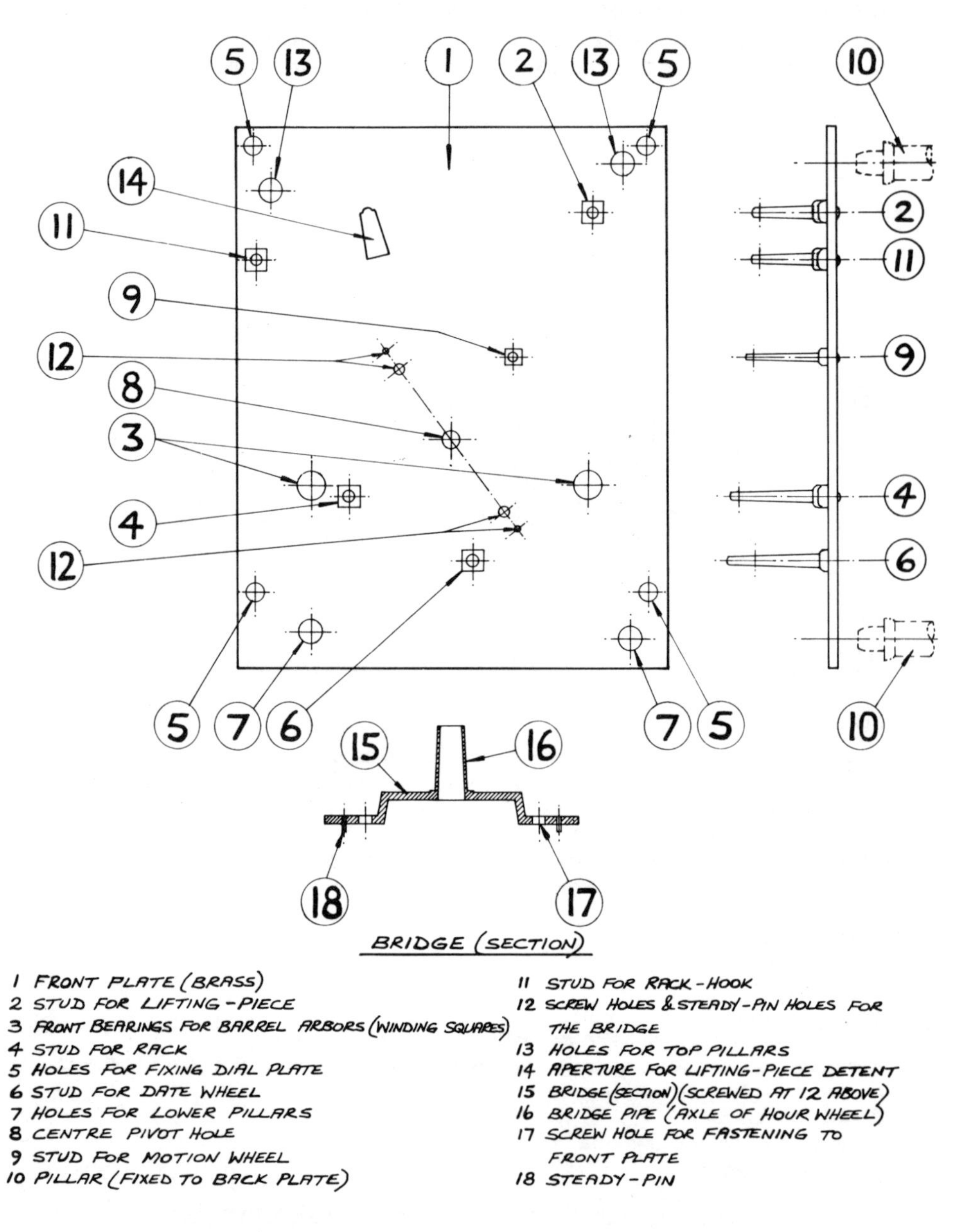

Figure 11

minute hand is manually turned against this friction, and of course it turns the motion wheel and the hour wheel too.

When the minute hand is put on the squared end of the pipe of the minute wheel, followed by a shaped washer called a hand collet, and the whole is pressed down against the friction pressure of the piece mentioned above, and fixed by a pin through a hole in the end of the arbor, the whole assembly is firm and tight.

The hour wheel turns concentrically with the minute wheel. It is accordingly mounted on a pipe fixed to the 'bridge' which spans the minute wheel. The bridge is pinned and screwed at each end to the front plate; and both the centre arbor and the minute-wheel pipe are contained within the larger pipe which carries the hour wheel. This larger pipe, however, does not anywhere touch the minute-wheel pipe (see Figure 9).

The bridge itself is not as wide as the diameter of the minute wheel, so that the motion wheel can be mounted to one side of the bridge directly engaged with the minute wheel. On the front surface of the motion wheel, about two or three millimetres from the rim of the wheel, is a short steel pin which protrudes forward at right angles to the plane of the wheel. This pin operates the strike mechanism, of which more later.

As we have seen, the hour wheel is made to turn on the pipe of the bridge, being driven by the six-leaved pinion of the motion wheel. It is thus turned concentrically with the centre arbor and the minute wheel, but at a much slower speed of course, because the hour hand moves only one-twelfth part of a circle while the minute hand makes a whole revolution.

Examine the hour-wheel unit, and you will see that it is — or should be — made up of five main parts. There is the centre pipe, the date pinion, the wheel itself, a snail-shaped cam (actually called the snail) and finally a large friction washer. It is not unusual for the hour-wheel hub and the date pinion to be all in one piece, the pinion being about one-third of the hub's length from its end.

Each side of the pinion is turned flat in the lathe, and the snail is fixed on its front face by screws or rivets. On the rear of the pinion, but free to rotate, is the hour wheel, positioned by the large-diameter friction washer which slides in a special groove cut in the pipe to receive it, in much the same manner as the washers on the great wheels of the clock.

The front end of the hour-wheel pipe varies in design, the pipe itself being circular in cross-section and thick, so that the forward end can be made with a shoulder, either square or circular according to whether the hour hand has a round or square hole in its boss. In both cases the hand may be held solely by a

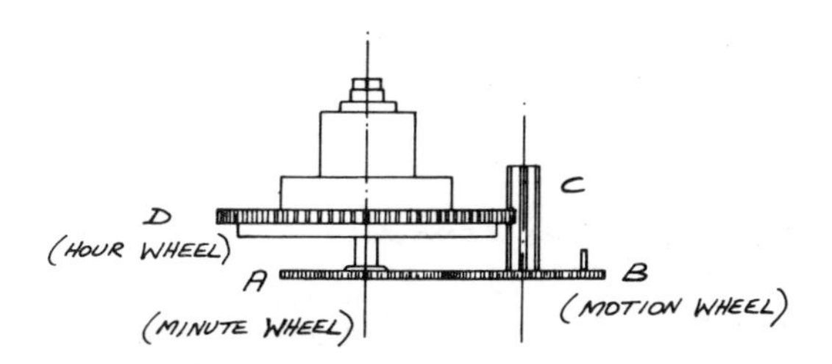

WHEELS A & B SAME SIZE, SAME NUMBER OF TEETH
PINION C HAS 6 LEAVES
WHEEL D HAS 72 TEETH

THE 'DRIVE' OF THE MOTION WORK
(SECTION 20)

Figure 12

tight friction fit or by a small screw or two inserted into tapped holes in the front edge of the pipe.

Where the hand has worked loose on the squared end of the pipe a fair repair can be made as described in Section 12, but study the diagrams in Figure 7.

Again, where an hour hand with a round hole in its boss has worked loose, a good fit may be restored by gently hammering the boss to reduce the size of the hole while the hand is held flat on a hard steel surface. This should be done carefully, with frequent tests to try the

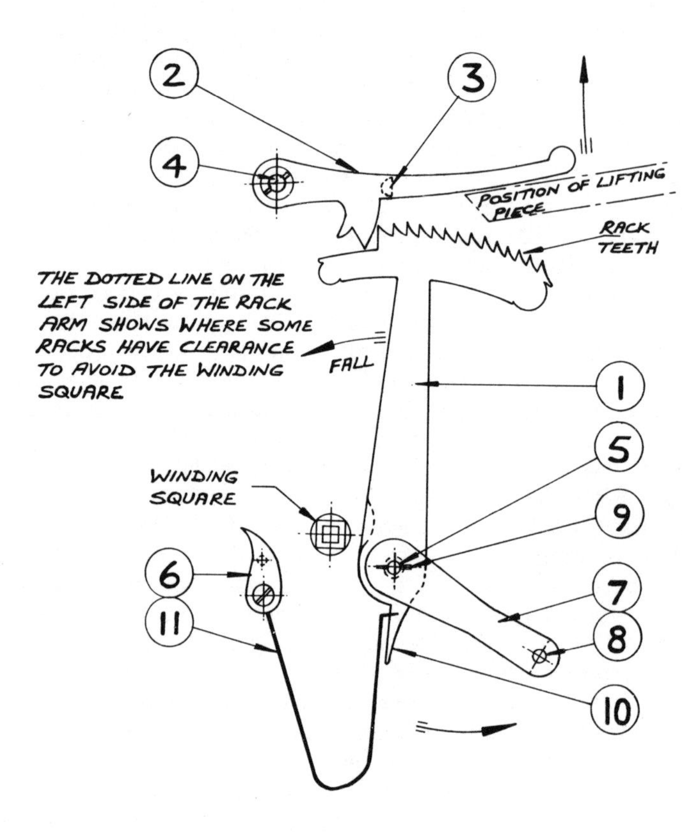

NOTE: FOR THE SAKE OF CLARITY THE GATHERING PALLET IS NOT SHOWN. IT WOULD BE BEHIND THE HOOK, ITS TAIL RESTING ON THE PIN 3. THE SKETCH BELOW SHOWS THE MORE USUAL TYPE, THE PIN BEING ON THE RACK ITSELF: IT IS THIS TYPE THAT IS DESCRIBED IN THE TEXT

IN THE DRAWING ON THE LEFT IT WILL BE NOTICED THAT THERE IS A STUD MOUNTED, NOT ON THE RACK, BUT ON THE BACK OF THE RACK HOOK. THE FINAL FALL OF THE RACK HOOK (INTO THE DEEP STEP OF THIS TYPE OF RACK) PLACES THE STUD IN THE PATH OF THE TAIL OF THE GATHERING PALLET AND SO LOCKS THE TRAIN

THE SKETCH BELOW SHOWS A MORE COMMON TYPE OF END TO THE RACK TEETH

1 RACK
2 RACK-HOOK
3 RACK-PIN
4 RACK-HOOK STUD
5 RACK STUD
6 RACK-SPRING MOUNTING
7 RACK TAIL
8 TAIL PIN
9 RACK RETAINER PIN
10 RACK EXTENSION FOR SPRING
11 WIRE SPRING

RACK, RACK-HOOK, AND RACK-SPRING (SECTION 20)

Figure 13

fit; if the beating is overdone, the metal becomes too thin – resulting, perhaps, in the necessity to solder a collar on the boss of the hand to make a 'push-on' fit on the hour-wheel pipe.

As regards the minute hand, care must again be taken not to beat the metal too thin and not to leave hammer marks. If the metal becomes too thin, the tightness of the oval friction washer on the square of the centre arbor – the one which drives the minute hand – will be affected.

Dismantling

After all that, let us return to the dismantling of the clock.

21 Having removed all the retaining pins, remove the rack, the rack hook, the rack spring and the lifting-piece. After ensuring that the very small pin (or sometimes a tiny nut) has been removed from the square end of the gathering-pallet arbor, gently lever up the gathering pallet to free it from the arbor. Place these parts in the bowl, or on one of the trays.

Sometimes the hour wheel and the date wheel have to be removed before the rack will be free, but in any case remove them – then the bridge, the minute wheel and washer, and the motion wheel.

When replacing pins it is always preferable to use new ones of the correct size, but if new ones are not to hand, old ones in reasonably good condition may be used again; they should therefore be cleaned, straightened and kept safely in the right order for ease in replacement.

When removing the date wheel, first make a small scribed mark across the edge of the snail on to the surface of the date wheel. This will be understood better with the wheel in front of you, when it will be seen that the date wheel is partly overlapped by the snail, being as it were sandwiched between – but clear of – the hour wheel and the snail. A mark made on the 1 o'clock step of the snail and continued over its edge on to the surface of the date wheel will give an exact line to which to work when the motion work is being reassembled (see Figure 14). The date ring shown overleaf is reproduced from an actual clock, but it could not have worked because the teeth are facing the wrong way. As these rings rotate anti-clockwise, the vertical sides of the teeth should be to the left.

The date wheel has either a steel pin or a brass 'finger' protruding from it which pushes on the date ring, one tooth daily. This wheel has twice the number of teeth that the date pinion has leaves; and as the latter turns twice in the 24 hours, the former turns but once. In consequence, the date changes very slowly, and is best brought about between the hours of 10 and 12 o'clock, or as soon thereafter as may be, so that when the clock is set up the change can be completed by midnight. If the change takes place at midday, move the hands on twelve hours, but if the clock has rack striking remember *never* to move the hands on beyond the 12 without letting the strike complete all twelve strokes. A count-wheel strike must, of course, be manually re-set.

22 Examine the fit of the gathering pallet on its shaft – there should be no play here at all. A very small amount of play may be cured temporarily by a touch of Locktite or similar preparation, but this is no proper cure; if you use it, take the greatest care not to let any of the fluid get into the pivot bearing.

If the clock has not been touched for some years it is virtually certain that the pallet pivot hole will have worn oval, in which case it should be re-bushed.

Before leaving the date work, it should be mentioned that on many clocks the twelfth step on the snail may be so positioned as to come on the same level as the teeth of the date pinion, ie be concentric with them. Where this is so, the front portion of the relevant teeth will have been cut away flush with the cam – but only the front part of the teeth.

A steel pin planted in the front plate of the clock to the left of the rack prevents the latter falling too far. The teeth of the date wheel run behind the cut-away portion and are therefore not affected.

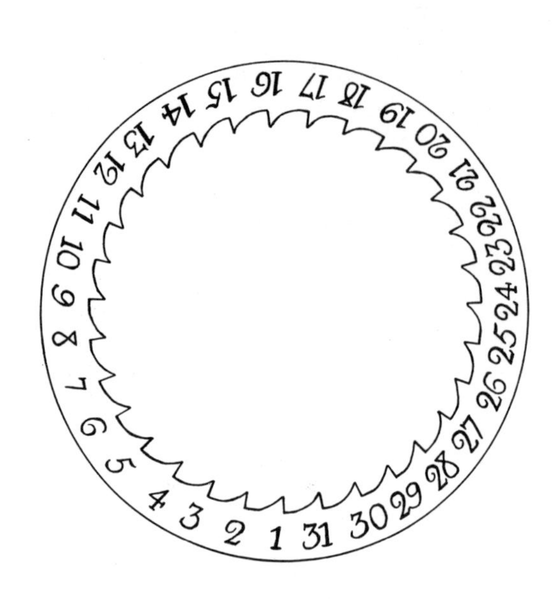

STRANGE DATE RING

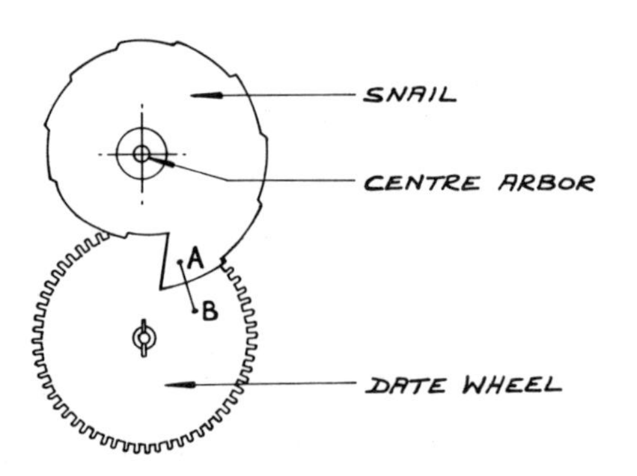

A LINE SCRIBED FROM A OVER THE STEP FOR '1 O'CLOCK' TO B ON THE DATE WHEEL WILL ENSURE REASSEMBLY AS BEFORE

DATE WHEEL AND SCRIBE MARK OVER THE SNAIL (SECTION 21)

Figure 14

23 At this stage you will have removed all you can from the clock without actually separating the plates. Before proceeding further, set the pallet wheel in the position it would be when at rest; and note the position of the pins on the pin wheel of the strike train. The hammer tail should be about halfway between two of the pins – there being eight in all.

(Note: on some longcase movements there is a metal 'stop' piece mounted on the top column by a screw. Its lower end acts as a spring stop to the rod which carries the hammer head. A certain amount of adjustment may be made by bending this in or out, thus varying the position of the hammer and its tail.)

It is essential when reassembling the wheels of the strike train to ensure that a certain amount of free run takes place before the pin on the pin wheel contacts the tail of the hammer; then the hammer will readily be lifted. If the hammer tail is fitted hard up against one of the pins, it nearly always happens that the clock will not strike.

When seeing that the hammer tail is between two of the pins, ensure also that the pin on the warning wheel is just above the detent of the lifting-piece. (The latter must be temporarily replaced on its post for this purpose.)

24 With the plates still in position, once again feel the play in the pivots of all the arbors, and note down any that are particularly slack. If the clock has not been touched for some years you will find that the front barrel arbors are slack, and probably both centre pivot holes as well.

The escape wheel and the pallet wheel, the fly and perhaps others may need attention; do not forget to examine the pivot hole in the back-cock which you have already removed. You will now have to do some re-bushing.

25 Sometimes the barrel arbors protrude behind the back-plate. Where this is so, to work on the movement without the cardboard box causes the arbors to be displaced. There are special movement-holders on the market, but for our use a box is quite satisfactory, especially as by rotating the box you can get at any part of the movement. Special holders are very useful when setting up the movement for testing after it has been repaired. To save expense you can make one for yourself, the method being described in the textbooks and magazines.

Separating the Plates

26 Now lay the clock on its back on the open box, with the open edges of the box supporting the movement. Choose a low seat by the bench so that your eye-level is not too high above the clock; arrange your bench lights so that they do not shine in your eyes. Use of the box prevents the arbors from being pushed up by the bench when the plates are separated.

Remove the hammer spring and replace its screw in its hole.

27 Count the pillars or columns to see if there are more than four, and remove the pins from them all. A pair of pliers is needed here, positioned so as to press on the thin end of the pin, and on the pillar stud *above* the thick end of the pin, and in line with the pin. Gentle pressure should cause the pin to slide out reasonably easily, but if it is tight a touch of penetrating oil should free it (see Figure 15). It is a mistake to force these pins in too tight, because doing so can split off the

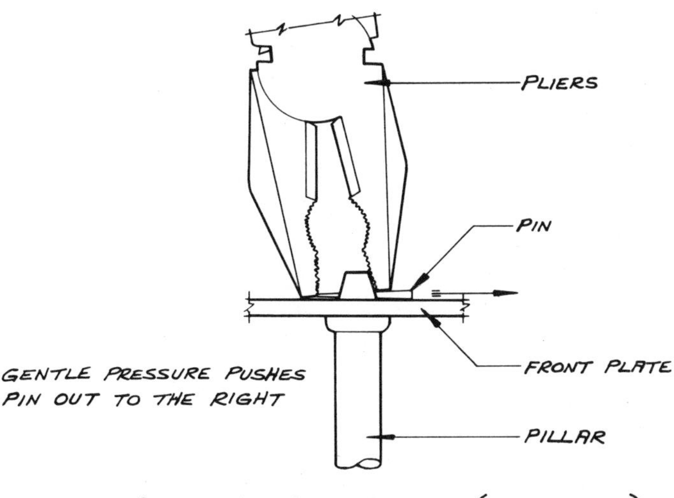

Figure 15

head of the pillar at the line of the pin hole. Where this has already happened, the repairer may have drilled a fresh hole in the edge of the plate through to the pillar hole and so into the pillar. If this is not possible, either a new pillar can be fitted or, more readily, the end of the pillar may be filed flat till its end lies *very slightly* below the surface of the plate.

The pillar end is then drilled and tapped to take a screw, which is provided with a slightly concave washer, put with its concave side downwards against the plate. The perfectionist will not like this because it is such an obvious repair; and the drilling of the hole in the edge of the plate is more in keeping with the kind of job a repairer might have done some 200 years ago.

Some plates on old clocks are latched. A small brass latch fixed by a rivet hinge to the plate slides into a slot cut in the head of the pillar. All latches should be slid back, of course, before the plates can be separated. As mentioned elsewhere, if the pins are rusted in or jammed tight, leave a drop of penetrating oil on the pin for a few minutes, and then with a fine hollow-ended punch lightly but sharply tap the pin out. If it breaks off, you may have to extract it by other means, even by drilling it out; if so, do not let the drill wander into the brass.

If the larger end of the pin is gently raised or very carefully levered up, it may be gripped with the pliers and so pulled out. Any levering must be *gentle*, for fear of cracking the top of the column.

28 We can now separate the plates of the clock. The columns or pillars are rigidly fixed to the back-plate, so that the front plate must be eased off upwards, the clock lying on its back on the box as described above. The front plate should be kept parallel to the back-plate all the time till it is clear. Using both hands, place the fingers beneath the edges of the front plate, and with your thumbs press gently down on the tips of the pillars, first at the base and then at the top, just enough to loosen the front plate.

Now ease the plate upwards, taking care that the cables do not pull the barrels out and upset the whole movement. The plate will first clear the fly, the warning wheel and possibly the hammer. You may now see why the hammer spring was removed earlier. Next the plate clears the ordinary wheels, then the pallet-wheel arbor, escape-wheel arbor, winding squares and finally the centre-wheel arbor.

29 Place the front plate on one side for attention later, and remove the wheels carefully to ensure that no pivots get broken or damaged.

Take out the barrels and keep each train separately for ease in reassembling, though this is very soon learnt. It is advisable to take out the barrels first. Then the wheels, etc, are removed and placed in the bowl. Then remove the cables from the barrels.

A strong piece of iron wire about 7 or 8in (200mm) long, with one end twisted into the smallest hook you can make, may be used as a tool to extract the line, which is held in the barrel by being threaded through an outer hole and knotted inside: the barrel, of course, is hollow. To enable you to get at this knot there is a small round hole in one of the end plates of the barrel. Hook out the line through this hole with the wire hook; the knot must be cut off to allow the line or cable to be pulled clear. Do not cut the line off outside the barrel because this leaves the knot loose inside. You may find that there are knots still left inside the barrel and all these should be removed. Holding the barrel with the hole downwards, pull the knots over the hole and then hook them out or extract them with fine-nosed pliers.

I once removed eight knots from the barrel of an old clock, any one of which could have caused much damage if it had fallen out while the clock was working.

Barrel Assemblies

30 Take up each barrel to examine the clicks or pawls on each great wheel, because any looseness in them will have to be remedied by soft hammer-blows on the outside while the click rests on a steel plate — as if tapping a soft rivet. Do not hammer the brass, as this will distort the wheel if overdone; and see that the click fits well into the ratchet teeth, both vertically and horizontally.

Adjust the tension of the click spring, which should press on the click near to its pivot, in order to minimise the travel of the spring. Some clicks have a tail which curls slightly upwards; these are preferable to those without tails, as the tail lies alongside the rim of the great wheel giving a supporting effect and preventing sideways movement. Any badly worn clicks should be replaced by new ones, which you may buy from materials dealers (if you have a workshop of your own you can make them, but it takes some time).

The click spring, generally made of brass, is screwed and pinned to the wheel, or fastened with two small rivets. If the spring is broken, a temporary one may be made out of brass wire, threaded into the rivet holes and hammered tight.

A new spring may be cut out of brass sheet, filed to shape and hammered to harden it, but this requires time and practice.

As yet, we have not removed the great wheel from the barrel.

Removal of the Great Wheel

31 Turn the barrel over to reveal the other side of the great wheel. The latter is positioned, but left free to rotate on its arbor, by a stout friction washer. This is slotted so as to slide into a groove turned in the arbor, after having been slid over the end of the arbor; a clearance hole is cut in the washer eccentrically, with a slot running into it, rather like a keyhole with straight sides. The slot is narrower than the arbor and will only fit into the groove in the latter.

32 This washer is usually held in place by a small pin driven through it opposite the slot, into the body of the wheel; and most of such pins are cut off so short that it is impossible to get them out – and exasperating to try. More modern types

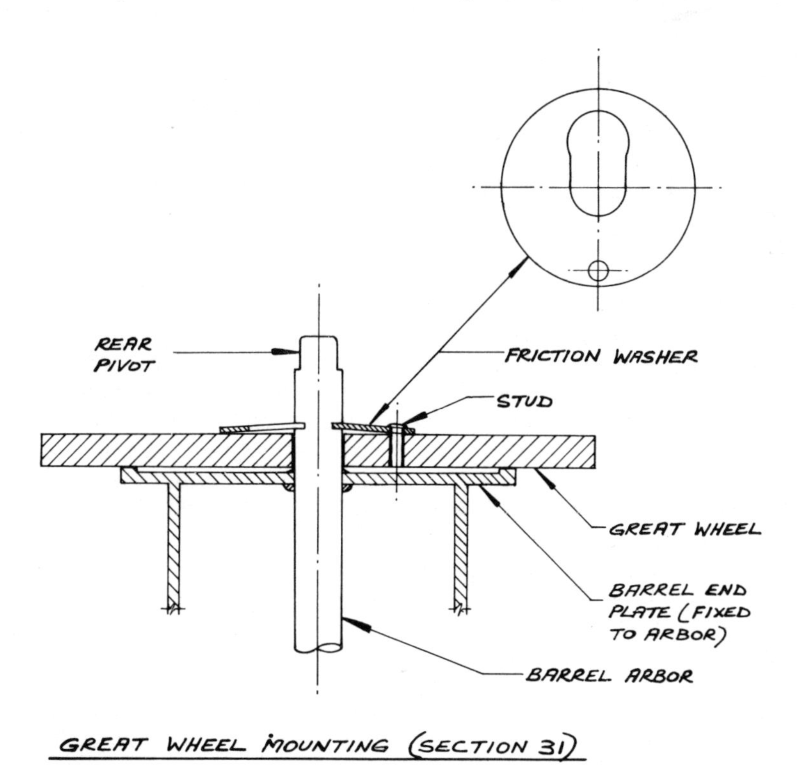

Figure 16

have small screws which present no difficulty.

The pins have to be removed, however, because otherwise the washer cannot be slid back to allow the wheel to be taken off. If the washer has to be lifted it may be damaged, so do not prise it up till you have a spare one to replace it. Now if the pin is filed flat and flush with the surface of the washer, the latter will not require lifting more than its own thickness; and the washer may then be pressed back over the end of the pin, which can be extracted later. If however you possess or can obtain an especially thin metal saw-blade, the washer can be prised up just enough to allow the entry of the saw, in order to cut the pin. When cutting the pin do not score the wheel. When the pin is cut, the washer may be slid back and the parts of the pin punched out of both wheel and washer.

If the washer is badly worn at the edges which engage the groove, or is otherwise badly damaged, you may be able to make a new one for yourself. Copy the old one as nearly as possible. After fitting it for size, shape and thickness, you will need to 'bow' it, that is make it slightly concave on one side and so convex on the other.

With a piercing saw, cut out a circular piece of brass slightly larger than the original washer and hammer it to harden it. On it mark the shape of the slot from the old one, drill a hole for the saw, insert the latter and cut out the 'keyhole'-shaped piece. While doing this the work should be held firmly on a fretsaw rest, ie a projecting wooden rest with a V-slot cut in it. The washer is best held down with an engineer's cramp, which can be moved from time to time as the work progresses.

Now cut a piece of ¾in (20mm) copper pipe about 1in (26mm) long, and stand it on end on the body of a vice, or on some solid hard surface. Place the washer centrally on this pipe and with the pean (ball-shaped) end of a mechanic's hammer, tap the centre once or twice. Test the washer on the wheel and adjust it to obtain a good friction fit which will not allow any lateral movement of the wheel. There should not be excessive friction between the barrel plate and the spokes or crossings of the wheel, and a touch of clock oil is needed here.

We are not concerned here with how the qualified and fully equipped engineer would do these jobs. We are trying to help you, as a handyman (and prevent irreparable damage to good clocks) by suggesting ways in which you may make reasonably accurate repairs to your grandfather clock!

When the washer is finished, fit it in place and mark its position on the wheel with a scribe. Dismantle it again and, without the arbor, re-position both the wheel and washer to the mark. Now mark the washer, this time through the hole in the wheel where the old holding-pin was.

Then, according to the size of the hole, either tap it with a 10 BA thread, or drill it to take another suitable tap as small as possible.

A cheese-head 10 BA screw (or other size, as the case may be) should then be fully screwed in, and cut off on the underside flush with the surface of the wheel. Now select a drill of a size slightly less in diameter than the *head* of the screw; and drill a hole in the washer at the point where it has been marked through the wheel. Remove the screw from the wheel, and broach out the hole in the washer till it is a good fit round the cheese-head of the screw. Assemble the wheel and washer on the barrel arbor and replace the screw. The screw head now acts as a stud or stop to prevent the washer moving out of position.

If the head of your screw pressed down on the washer when tightened, it would set up a stress on the bowed washer, and so lessen its friction grip, which is bad. In the course of many decades of winding of the clock, there will be wear on the slotted edges of the washer in which the arbor rotates during winding. Slight pressure will be maintained where there is no screw pressure, and the wear will virtually cease when the fit is ideal.

It will readily be realised that if the great wheel does become loose, no matter how slightly, its centre hole — which is a true fit on the arbor — will become worn; and when this happens the centre pinion

Using a small home-made tool to extract the hand-retaining pin from the end of the centre arbor, before cleaning the circular silvered brass dial (Section 10)

Twelve-inch square brass dial with silvered chapter ring, date and seconds rings, nicely matted centre, and a bevel on the edge of the date aperture for easy changing. James (or John) Foy, Taunton, c1780 (Section 14)

An ordinary break-arch painted dial with an arc date aperture, brass rings to the winding holes, and no minute numbers. It has opposed spandrel pictures, and a picture in the arch which is badly out of proportion (Section 14)

Depth tool showing lantern runner with mounted escapement. The crutch is free to oscillate (Section 61)

begins to wear because the depthing is affected, and trouble ensues all along the line.

33 Before we leave the great-wheel assembly, the journals (the parts of the arbor which turn in the bearing) should be examined. Normally, if these are well made, hard and polished, nothing more than a good clean is needed. See to the ratchet teeth on the rear end of the barrel, and true up any damaged ones, filing the sloping sides (only) to achieve a good fit on the click.

Make sure that the click enters the space between the teeth and fills it fully, as this will prevent uneven wear. Lightly oil the clicks with clock oil and wipe off any surplus oil.

34 Reassemble the great wheel, etc, on the barrel, and place the latter in its pivot holes in the plates, which should be lightly pinned together — one pin in each of two opposite corners will do — and test the run of the arbor. There should be fully free rotation, no lateral play, and only moderate end play. Quite a good test for end play is to tilt the plates over and back while watching each wheel movement.

Every arbor should slide under its own weight to the lower plate without excessive play. Excessive play may be corrected by inserting a thicker bush in the back-plate (the back-plate is preferable because it is difficult to assemble the arbor in the front plate — it cannot be slid into its hole, see Figure 17).

A little end play in the case of the great wheel is no bad thing as it tends to spread the load on the centre pinion. If no play exists the wheel-teeth fall on the same spot on the pinion leaves each time, and will eventually wear a sharp groove.

For the sake of this exercise, let us assume that there is side play on the bearings. First, make certain that the barrel arbors are true. You have either trued them in the lathe or got an engineer to do it for you. You are now faced with two jobs: a) to re-bush the bearings and b) to check the depthing of the great-wheel teeth with the centre-wheel pinion, remembering that you must not move the position of either of the *centre* pivot holes.

Therefore the exact position of the new bushes will depend upon proper depthing; but this is done in a special tool, which you will not have. The process, however, must be touched upon: in fact, the more experienced repairer will be able to make a depthing tool. The idea is to mount the wheel and pinion in a tool capable of being finely adjusted to achieve a good free run with the least possible friction between wheel and pinion; it should be fitted so that the centres of the two arbors can be accurately measured or pin-pointed on to the plates being scribed as when using a compass. Then the holes drilled in the plates in the positions so marked will be

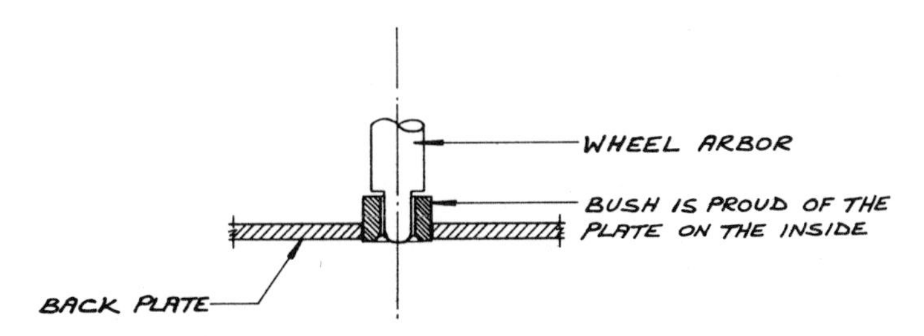

METHOD OF TAKING UP END-PLAY BY A BUSH FITTED PROUD OF THE PLATE (SECTION 34)

Figure 17

exactly right for the two arbors concerned – ie the depthing will be correct. You can readily buy a depthing tool, but being precision instruments, they are expensive – from £65 to £125 at early 1980s prices. (See also Section 61; Plates 8, 9 and 10; and Reference Sources page 96.)

35 Now the journals – the great-wheel pivots – have been slightly reduced by being trued in the lathe, and the holes in the plates, being worn, will not only be larger but will be oval in shape, even if only slightly. This oval wear is caused by thrust of the great wheel away from the centre pinion. Close examination will reveal that both centre hole and barrel hole are worn on sides opposite to one another.

It is axiomatic, of course, that these holes must be made circular, and in such a way that the centre of each circle is the exact distance as measured by the depther after testing the engagement of

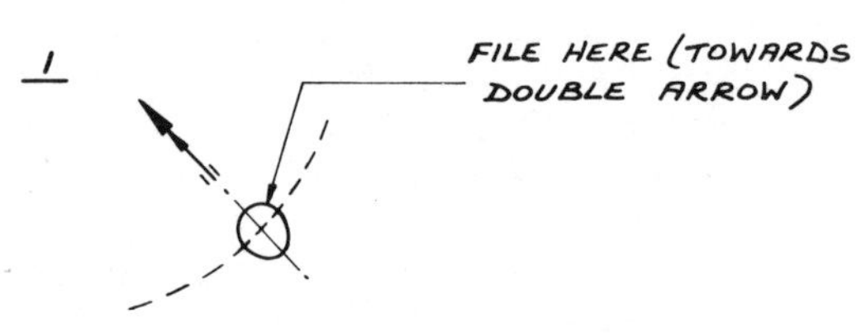

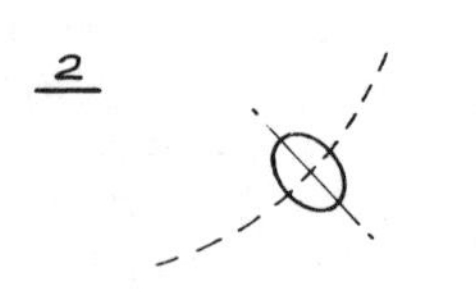

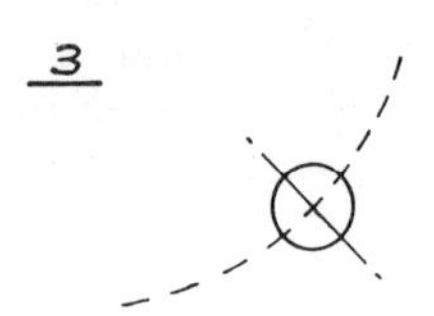

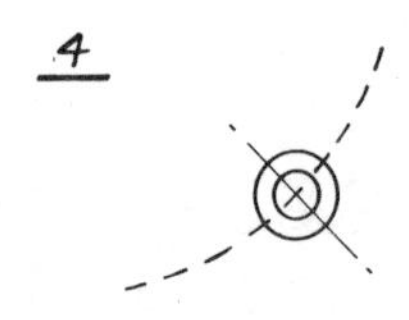

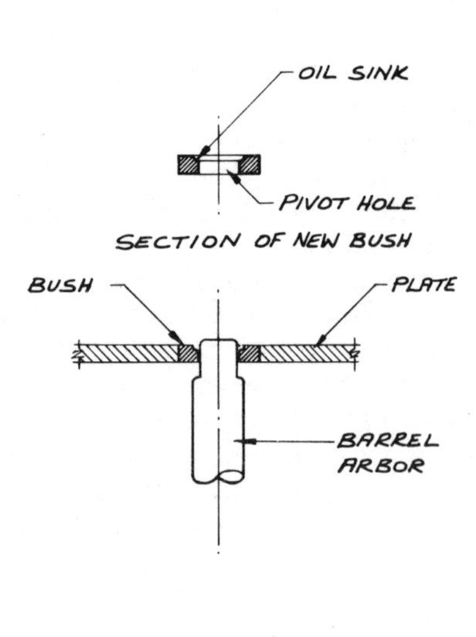

Figure 18

wheel and pinion. When using a proper depther, to mark this distance one point of the tool must be placed in the centre pivot hole on the inside of the plate, and the other point used to make a scribe mark across the barrel-arbor hole. If this scribed arc can be judged to pass through the centre (unlikely), the hole can then be broached circular with a suitable 5 or 6 sided steel broach. (You cannot do it with a drill.)

Assuming, however, as is probable, that the wear caused by thrust is at a point away from the centre hole, then more of the hole, so to speak, will be on the far side of the scribe mark than on the near side; so the latter will need filing gently till the hole is central, balanced about the scribe mark.

This is called 'drawing' a pivot hole, and at this stage it will not be circular but oval. Having done the filing, now make the hole circular with a suitable steel-cutting broach, taking off only as much metal as is strictly necessary (see Figure 18).

Broach only to make a *tight* fit for the new bush; and if the hole is very much worn, it is better to file and broach as you go. This is much easier on the broach and results in a more accurate hole. In this way you will make the hole ready to receive a new bush.

But you will find that the broach will have 'burred' up the metal on the edges of the hole. Sometimes this burr may be left and later hammered flat to hold the new bush in place. Alternatively — and preferably — if the bush is slightly thicker than the clock plate, the hole may be gently counter-sunk on both sides, and the new bush carefully rivetted into the plate. This will, of course, distort the hole for the pivot, and the former should be broached out to make a good fit, not forgetting that the broach must at all times be kept at right angles to the plate. Here again it should be said that there are special tools for doing this job, but the skilled amateur who observes the rules can do quite well without them.

In describing the above it has been assumed that the clock's centre pivot hole is all right. If it is worn, it should *first* be re-bushed in order to get the distance exactly correct for the larger hole.

In all cases it is essential to insert the bush so that its hole is truly at right angles to the plate.

36 New bushes or bouchons can be obtained from any dealer in such supplies, in boxes of varying sizes and thicknesses, and a clockmaker will keep a stock of them. As you are not a clockmaker yet, and do not want to buy in quantity anyway, the best thing is to buy some brass rod as you require it.

In the lathe, turn a piece of brass rod to fit the hole tightly enough to be pressed in with the fingers. Cut it off to the thickness of the plate plus a little extra. Now place the piece in the prepared hole, and on a flat steel surface hammer it flat on both sides. Do not use heavy blows, but sufficient to flatten the burrs on the edges of the hole to grip the new piece tightly.

Now scribe the arc of distance across the new piece, and where it crosses the middle, punch a mark with a centre-punch to show where you will drill the hole. Take special care in doing this and make sure that the punch mark is exactly on the arc and as central as possible.

Now drill a guide hole exactly at right angles to the plate and widen it out till it is very nearly large enough to accept the pivot. From the *inside* broach the new hole till you can press the pivot in by hand. Withdraw it, and with a smooth round broach, again from the inside, broach the hole in the direction of rotation of the arbor, using a light oil to ease the broach, which again must always be kept at right angles to the plate.

Having done the back-plate, you will have to turn to the front plate, but you will see that the front pivot hole of the barrel arbor is very large, the largest of all in fact. It is therefore essential in this case to buy a special bush, and *not* to plug the hole and re-drill as described above.

If the wear at the front pivot hole is only slight, and if the burr of previous broaching is still in evidence, the hole may be made smaller by gentle hammering on a flat steel plate; but care should be taken not to mark or dent the brass, ie make sure that the hammer face is parallel to the steel plate. Broach only with a smooth round broach in this case, and continue carefully till you have obtained a good fit to the arbor.

The next stage is to insert the arbor in the plates, to pin the latter lightly, or hold them firmly, and to try the run. If it is very stiff put a little oil on the pivots. If this makes no improvement, find out where the resistance occurs, and very carefully broach the relevant hole with the round broach, using a little oil and more pressure than before, and try again. When the arbor spins nicely in its bearings, this part of the job is complete.

Try the end-shake to see if it is free but not excessively so, because sometimes at this stage the arbor 'sticks', on one side or the other; if it does, remove it from the plates, and with a wide counter-sink remove the extreme edge of the *inside* of the new bush or bushes. This gives a small amount of end-shake, but take care not to overdo it.

Now put the barrel arbor complete with great wheel, etc, and also the centre wheel, in the plates and try the depth. Test by moving the wheel very slowly to see that the run is free all the way round the great wheel.

Depthing (General)

37 To ensure that you can master the depthing, you need to understand something about gearing.

Look at a toothed wheel, and imagine two circles, one with a radius from the base of the teeth to the wheel centre, and the other from the tip of the teeth to the same centre. The difference between the two radii is, of course, the height of the teeth. If you examine a wheel, it will be seen that the teeth are all pointed like an arch, and that there is quite a space between each. It is into this space that a leaf of the pinion will fit, and it must therefore be of the correct size.

The pinion leaves, on the other hand, are rounded, being thicker on the outside than they are nearer the centre.

If a line is drawn, as it were, joining the centres of two wheels, or in the case of clocks joining the centres of a wheel and a pinion, that line is called the line of centres. The teeth of the wheel 'drive' the leaves of the pinion. Now, the best condition of contact between tooth and leaf is very slightly after the line of centres; but this requires multi-leaved pinions. A wheel of, say, 60 teeth would require a pinion of not less than 11 leaves if the driving thrust is to be after the line of centres.

Such pinions are seldom found in the ordinary longcase clock, which has pinions of 6 to 10 leaves, averaging 7 and 8; but they work perfectly satisfactorily and efficiently. The reason why the leaves of a pinion are rounded is to obtain action as near to the line of centres as possible, so it is always best to ensure that the pinion leaves are highly polished to reduce friction.

If you have done your depthing correctly the pinion leaves between the wheel-teeth will be neither too deep nor too shallow. If they are too shallow, excessive play will be readily apparent; but if they are too deep you may get what is called 'butting'. A simple analogy can best describe the effect of this. Assume that you have an umbrella which you are dragging point-downwards along the pavement in the street. The umbrella offers no resistance and is not in any way impeded by the cracks between the paving stones. Now try pushing the umbrella in front of you, and feel the difference. It offers resistance all the time and will stop dead at the first crack between the slabs.

Thus, if the pinion leaves are set too deep there is not enough clearance for them to clear the wheel-teeth. Try this for yourself in the depther; a small mirror held at a suitable angle will enable you to see the engagement of the teeth better. Rotate the wheel very slowly. While one tooth is turning a pinion leaf, the tooth behind it is entering the next slot, as it were, but if it touches the tooth behind that again you get butting (pushing the umbrella), and the clock will stop. Never try to get over this by oiling the pinion, because this will do more harm than good. Never oil wheels or pinions, only the pivots and anchor pallets — more of which later.

The pinion leaf receives power from the wheel-tooth by pressure till the point of the tooth approaches the end of its contact with the pinion leaf. At this moment the next tooth behind it is just about to engage the next pinion leaf, and so on. One cannot adequately describe this on paper without going into technicalities, mathematics and geometric designs. Here we are just dealing as simply as possible with a grandfather clock; you will know for yourself whether any job that needs doing is beyond you, and if it is, then you should of course take it to the professional man. Remember that your clock is worth quite a lot — anything from £200 to perhaps £25,000, and it is wise to get work properly done.

38 Complete the depthing of the other wheels of the going train and any re-bushing that is necessary. The escape wheel may need attention to its pivot, which should have no side play at all: any looseness here means a serious loss of power — the power which restores momentum to the oscillating pendulum.

We will return to the escape wheel later, but meantime let us turn to the strike train. The re-bushing process will have to be repeated here for all the pivots that need it; and when the re-bushing is completed, re-form the oil sinks on the outside of each pivot hole.

An oil sink is a circular depression drilled on the outside of all the smaller pivot holes to form a sharp circular edge with the surface of the plate. Oil put in a pivot hole which has no sink tends to creep away, leaving the pivot dry. Some makers — not all — hold that an oil sink, if well made, prevents this, and will hold a miniature oil supply to the pivot for a long time, in ideal conditions. But of course the real enemy here is dust. With the large pivot holes, eg those of the barrel arbors — but only with the large holes — it is not a bad thing to use a slightly thicker or heavier oil, because it is these pivots that carry the weights of the clock. Too light an oil is rapidly squeezed out.

39 When you have done all that is needed to the strike train, reassemble it, including the bell-hammer and spring, leaving out the going train with the exception of the centre wheel (see below). Assemble bridge, rack, rack hook, lifting-piece, rack spring and the entire motion work, all with the object of getting the wheels in their correct positions.

You will remember that the gathering pallet is squared on to its arbor, and can thus be fitted in only one of four positions relative to the wheel. Its 'stop' position is roughly horizontal with its tail towards the left — your left as you face the clock.

It is held in this position by the stop-pin on the rack, which in turn is located by the rack hook. With these pieces in place, the position of the tail of the bell-hammer is important. It should just have fallen off one of the eight pins of the pin wheel, coming to rest approximately halfway between two pins. To observe how all the parts operate, it can readily be understood that you must assemble the bridge, minute wheel and motion wheel on the front plate; and this, of course, is the reason why we left the centre wheel in place.

40 When mounting the motion wheel proceed as follows:

a) Place the minute hand on the square of the minute-wheel pipe and point it to 12 o'clock.

b) Place the motion wheel in position so that its pin has just — but only just — passed the lifting-piece tail, which drops. Adjust the engagement of the two wheels so that the hand points directly upwards (in the 12 o'clock position) when the 'drop' of the lifting-piece occurs.

c) Holding these two wheels in this position, adjust the engagement of the hour wheel so that the rack-tail pin will fall midway on to the twelfth step of the snail.

d) Pin these parts in position for the moment, and now try out the action of the strike.

With the movement standing upright on the cardboard box, rotate the minute hand clockwise to about the 9 o'clock mark, and then continue slowly, at the same time pressing the strike great wheel down in its direction of rotation (anti-clockwise). As the hand comes to the 10-minute position or thereabouts, the pin on the motion wheel will start to press on the tail of the lifting-piece, raising the latter which soon comes into contact with the arm of the rack hook.

At about 5 minutes to the hour the rack hook will be lifted clear of the rack, which will 'fall' under the pressure of the rack spring, and at the same time the gathering pallet will be released.

This allows the pallet wheel to rotate and, if there was no mechanism to stop it, the clock would strike. But there *is* such a mechanism: on the lifting-piece is a flat steel blade set at an angle to the vertical, and protruding from the lifting-piece inwards through a hole in the front plate. When the lifting-piece is raised – as it now is in our description – this blade is in the path of the pin on the warning wheel.

Thus it is very important that this pin should be caught by the blade *before* the wheel has had time to rotate very far. If for instance the warning wheel turned almost a full revolution before its pin was caught, the result might well be that the train would run far enough to bring the next pin on the pin wheel hard up against the tail of the bell-hammer. In this position the clock would not always strike.

This action depends on whether the blade of the lifting-piece, called a detent, has risen far enough to place itself in the path of the warning-wheel pin *before* the rack falls. It follows that if the rack falls too quickly the action may be nullified. Therefore it is quite important to set a clearance between the lifting-piece and the arm of the rack hook. This arm can be bent (under heat) quite easily to achieve the clearance. The maximum lift should ensure that the tip of the rack hook clears *all* the teeth of the rack.

Strike Action

41 The sequence of events is as follows:

a) At about 10 minutes to the hour the pin on the motion wheel contacts the tail of the lifting-piece, which starts to lift.

b) A minute or two later the upper arm of the lifting-piece contacts the underside of the arm of the rack hook, which is thus lifted.

c) The point, or hook, of the rack hook begins to rise upwards out of engagement with the (first) tooth of the rack. (Note: on some clocks the rack hook is not in the first tooth, but in the second or even third.)

d) In the next minute or so the detent is in the path of the pin on the warning wheel, and the rack is released.

e) The release of the rack lets it fall, being lightly pressured to do this by its spring. This action also releases the tail of the gathering pallet; that is to say the pin on the rack is carried away to the left as the rack falls, and because the tail of the gathering pallet was resting on the pin it is now freed.

f) The pallet wheel turns, causing the warning wheel (and fly) to do likewise till its pin is caught by the detent on the lifting-piece.

g) The clock is now ready to strike, but cannot do so because the lifting-piece is still held up by the motion-wheel pin.

h) The setting of the motion wheel is such, as we have seen, that when the minute hand reaches the 60-minute mark on the clock dial, the tail of the lifting-piece falls off the motion-wheel pin, and the detent is released.

i) The clock strikes the correct number of strokes as determined by the position of the rack tail on the snail when the rack falls.

In the example given above it would strike 1 o'clock (see Section 40).

Rack and Gathering Pallet

42 It will have been seen that the lifting-piece falls by gravity. There is a slight pressure downwards from the warning-wheel pin which assists the fall.

The rack, on the other hand, has a spring to ensure its action, and the reason is obvious: the rack is almost vertical and is thus nearly, but not quite, in balance. Furthermore it is under slight 'grip' pressure from the tail of the gathering pallet, and the direction of this pressure is of some importance. Remember that the rack hook locks the rack, ie prevents it from falling to your left as you face the clock. It will not, however, stop the rack from moving to the right, so that the pressure of the tail of the gathering pallet must never tend to push the rack to the right.

It is hardly necessary to mention this, but knowing it may save you trouble later on. Thus if we go to the other extreme and make the tail of the gathering pallet press the rack pin hard to the left, it might well cause the rack hook to jump out of its rack tooth and so release the strike. Midway between these two extremes is the ideal position to find, and this may be done by drawing an imaginary line, so to speak, through the centre of the rack pivot and the centre of the rack pin.

Now if the pressure on the pin is straight down this line to the rack centre, no force will be exerted either to right or left.

If the pallet tail is slightly bevelled or rounded so as to exert a slight pressure to the left of this imaginary line, then when the rack hook lifts, the pallet will help the rack on its way. At the end of a 'strike' the momentary greater pressure of the pallet tail being caught by the pin on the rack is not enough to force the rack hook out of its tooth, and thus the train is satisfactorily locked.

Lifting-Piece

43 Before leaving this subject it is as well to refer to the lifting-piece once more. Sometimes the detent catches the

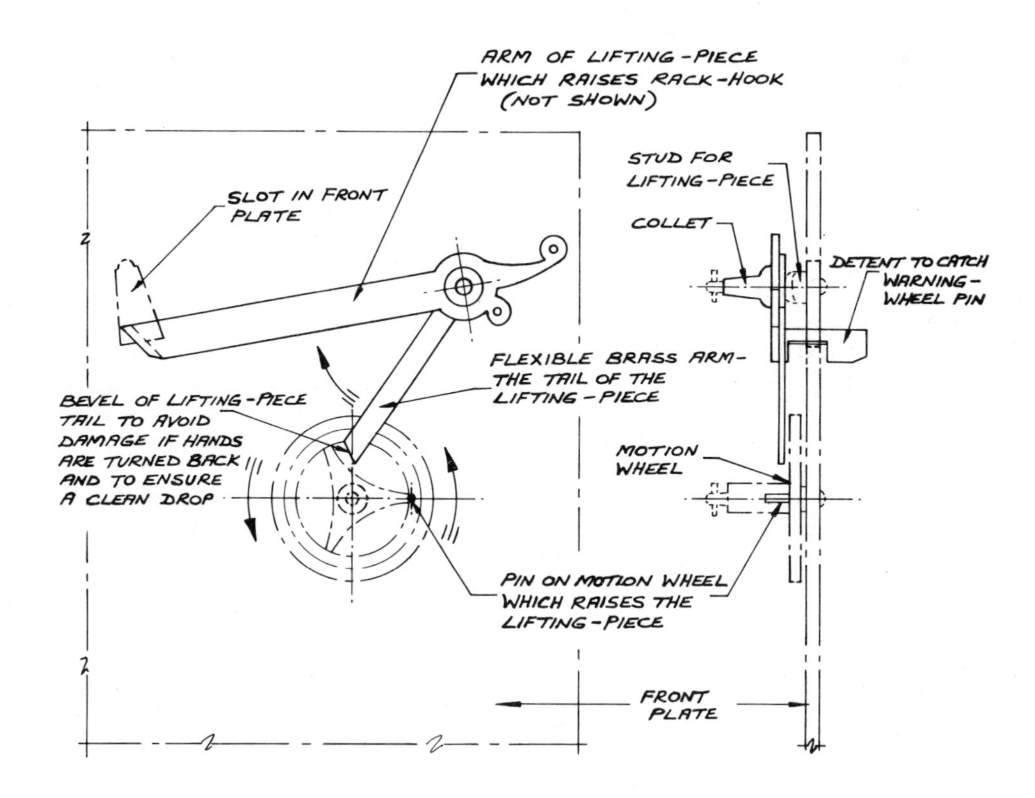

Figure 19

warning wheel satisfactorily but continues to lift after so doing. This is bad, because it turns the warning wheel backwards — no matter how little — and the fly will also turn back, perhaps half a turn. So, in theory at any rate, the whole train is turned back, and the weight of the strike train actually raised, albeit by an infinitesimal amount. In practice it is only the 'slack' or play in the train that is taken up; and, of course, there is increased friction.

From the point of view of mechanical advantage, the tail of the lifting-piece should be a shade longer than the detent arm to obtain good leverage, in order to lift the rack hook out of engagement with the rack. In practice the lifting-piece tail is often quite short; where this is so, the extension or arm of the rack hook will be long, and will rest on the detent arm of the lifting-piece about halfway up towards the pivot stud. The result of this is that a good sound leverage is acquired without affecting the going of the clock.

Study the teeth of the rack. Their left side is vertical and as the right side of the rack hook is also vertical, or very nearly so, the locking of the rack is secure. The right side of each tooth is sloping, so that movement of the rack to the right by the gathering pallet easily slides the hook into the next tooth back. The lifting-piece literally lifts the hook upwards and has only to overcome the slight frictional contact between the two surfaces as described above.

Once the pin on the motion wheel has caused the lift, the work has been done, and it only remains for the pin to hold up the lifting-piece *until the time comes for it to fall.*

Viewed from the front, the motion wheel rotates anti-clockwise, and its pin pushes the tail of the lifting-piece either upwards or to the left, according to the design of the lifting-piece. Where the tail is pushed upwards, contact is made with the pin when it is, so to speak, at about 2 o'clock on the face of the wheel; and lift proceeds till the pin reaches top dead centre. Thereafter a drop begins.

Where the tail is pushed sideways, contact is made about 10 o'clock on the wheel and drop occurs at 9 or 8 o'clock. Once the rack has fallen, the pin on the motion wheel should exert no further lift, but it should hold the lifting-piece up until the minute hand of the clock reaches the 60-minute mark on the dial. If we shape the tail of the lifting-piece to allow for this, we can prevent further lift so that the train will not be pushed backwards as described above.

It sometimes happens that when setting the point of 'drop off', when the tail drops to allow the clock to strike, the minute hand is just before or just after the 60 mark on the dial. Final and exact adjustment can generally be made by slightly bending the motion-wheel pin till the desired effect is achieved. The dial must, of course, be removed to do this.

Pivots

44 A word now about pivots. After cleaning out the pivot holes with pegwood (cocktail sticks will do if you have no pegwood such as clockmakers use), and after washing and cleaning all pivots, fit each wheel in turn, alone by itself, in the plates.

Any undue play in the pivots at once shows itself, but before any re-bushing is done it is essential to see that the pivots are quite true. There is no simple way of achieving this and you are not advised to attempt the truing of pivots unless you know just what you are doing. You have been warned!

If you look at a worn pivot under a lens, you will find two main faults: the pivot has worn to a dumb-bell shape; and the steel has worn into a series of rings or tiny grooves. The ideal pivot should be truly cylindrical, hard and polished, and terminate in a small cone or hemisphere, ie with a true centre.

Those found in a grandfather clock are comparatively large and can be trued in a lathe. Some pivots are slightly tapered, thicker at their inner ends and thinner towards the tip. This is logical where the bearing or pivot hole is shaped by a tapered broach, but the objection to it is that any end-shake tends to 'grip' the pivot; and power applied sideways to such a pivot must necessarily set up end-shake by ordinary 'wedge' action — no matter how slight.

If wear is excessive, a new pivot will have to be made and fitted, a job well beyond the skill of the ordinary person. Even a skilled turner needs to take concentrated care in doing it, in order to end up with a dead true, hardened and polished pivot.

With proper lathe attachments, pivots can be safely replaced by drilling out the arbor, having cut off the old pivot, and inserting the annealed piece of an ordinary needle which has been very slightly end-tapered, suitably cut, shaped and finally re-hardened. The arbor is drilled out, carefully reamed, heated and shrunk on to the piece of needle.

This is a skilled job, as you may see, and while you may be able to overcome the practical side, could you be sure that your pivot is dead central? If not, do not try it, but take the wheel to a clockmaker or expert model-maker.

There is a tool for polishing pivots called the Jacot tool, but this again is not for the amateur. If however you are an ingenious and careful handyman, you may proceed as follows:

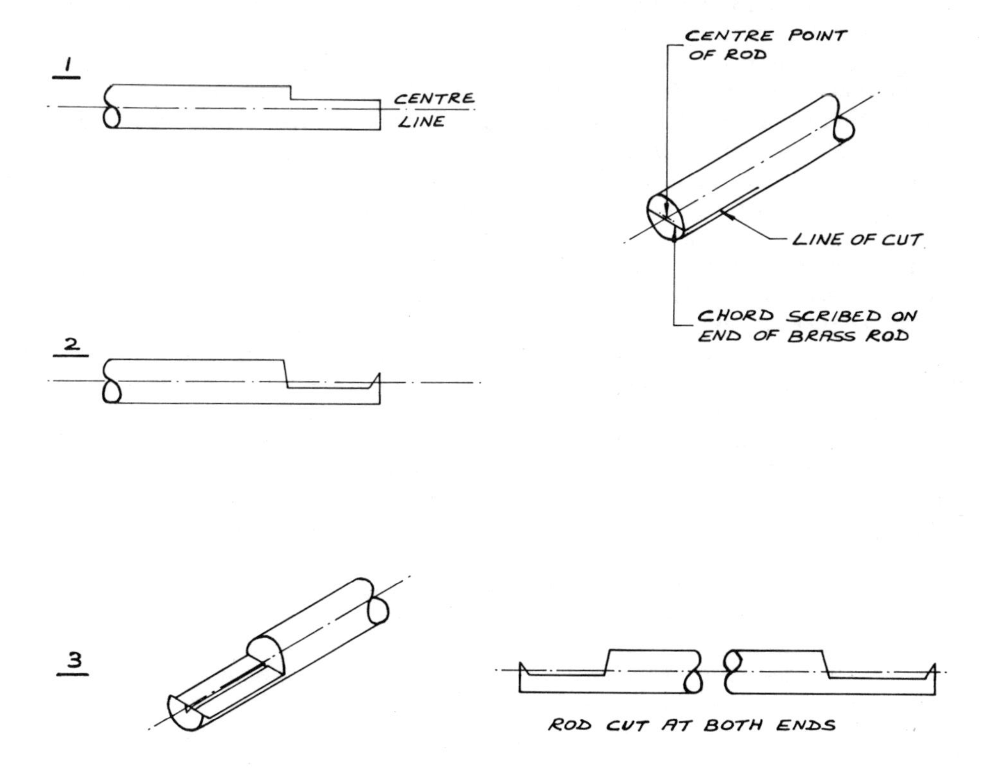

Figure 20

You need two bench vices, a hand drill, some tools and a small hand vice. Take a ¼in (7mm) diameter brass rod, see that one end is flat and true, and on that end scribe a chord about 1mm from, and parallel to, the diameter. Fix the rod upright in a vice and cut downwards on this line for about 15mm and then cut out the thinner piece. Use a fine blade saw for this and keep it parallel to the diameter all the time. This operation is best understood by examining Figure 20. You will now have a piece as shown in 1 – in the side views the centre of the rod is indicated by a dotted line.

Now file out a piece within the cut portion in the manner shown in 2, and it will be seen that the bed of the filed-out part is just below the centre line, and that a raised portion is left at the end of the rod. Exactly at the centre of this a V groove must be cut (filed) so that the end will look like 3. For the larger pivots a U-shaped groove is suitable.

Fix the wheel arbor in the chuck of the drill, and the latter in a vice so that the chuck can be rotated by hand. Fix the brass rod in the hand vice and the latter in the second bench vice, adjusting both so that the relevant pivot rests firmly in the slot in the brass rod. The rod must be secure and firm, and the pivot resting in the slot must also lie on the bed of the cut-away portion.

Now with a flat piece of oiled Arkansas stone held in one hand, and the arbor rotated by the other under light pressure, gently reciprocate the stone so as to true the pivot. Never use the stone on the pivot while the latter is still, because this will cause a 'flat' and ruin the pivot.

Before doing work on an actual pivot, practice many times on odd bits of steel rod or wire till you achieve success. When truing a rear centre-wheel pivot, never leave it too thin because it may break if left weak.

The above will horrify a professional clock repairer; nevertheless it is better than putting a worn pivot back into a new bush in the plate of a clock. But it is no use attempting it unless you are experienced in the use of ordinary mechanical tools.

In passing, it may be mentioned that several bits of brass rod may be cut for different thicknesses of pivot, and that both ends of such pieces may be cut. If you have a lathe, the brass rods are made to fit as runners in the tail stock.

Cleaning and Pegging Out

45 Returning to our clock, we now assume that we have cleaned and dried all the parts, trued the pivots that needed it, and completed all the re-bushing. The plates themselves and the various brass attachments such as the bridge and back-cock will have been cleaned, washed and dried, and the pivot holes pegged out. Keep pegwood pointed by scraping with a sharp knife after each use – insert it in a pivot hole, rotate it and withdraw it. It will have a black ring on it. Scrape this off till the wood is white again, and repeat until the wood remains clean when withdrawn.

Now, with its columns upwards, place the back-plate on the cardboard box as before, and replace all the wheels, the fly and the bell-hammer. It is wise to leave out the hammer spring at this stage. Fit the cables in their barrels, taking care to position them correctly: the going-barrel cable between the two bottom pillars, and the strike-barrel cable outside to the left, but of course below the arbor of the hammer. Check the positions of the warning wheel and the pin wheel, as already described.

In reassembling any clock, a strong pair of tweezers about 5in (130mm) long will be found useful, if not essential, and if they are cranked so much the better.

Replacing the Front Plate

46 Carefully hold up the front plate, right way up, above the assembled wheels and lower it gently, easing it first over the centre arbor, then over the winding squares, the escape-wheel arbor, and the arbor of the pallet wheel. Ease the plate down without any force, fitting it over the bottom pillars until you can lightly push the two bottom pillar-pins into, but not through, their pillar holes.

Placing the pins loosely thus will prevent the plate rising, and may even save breaking pivots. See that the plate is not caught up on the shoulder of the centre-wheel arbor.

Re-inserting the Pivots

47 By gentle trial and error, turning the clock round, ease each pivot into its place, leaving the fly till last; in fact sometimes it can be put in afterwards if mounted very near the edge of the plates.

See that the pivot of the warning wheel does not get into the clearance hole for the detent of the lifting-piece, for if it does, you will have to raise the front plate again in order to get it out. It may also indicate that the rear pivot of the warning wheel is too slack – not well re-bushed.

See that the hammer arbor is placed correctly – it is best to leave the hammer spring out for the moment and fit it in afterwards, because if it is in place when you are fitting the plates together, it may push the hammer out.

Remember that the taper on the column pins should only be enough to prevent the pin from passing right through the hole. If the pin is too tapered or conical, the vibration caused by the striking of the clock may well result in its dropping out. Such a pin would in theory be touching in a 'circle' of contact only, and even a slight tap would loosen it.

When the front plate is fully down and seated squarely on the pillars with all pivots free, replace all the pillar-pins (or latches as the case may be). Always fit new pins if the old ones are in any way damaged or suspect; and new pins should be tested in their holes before assembly, because a pin which fits a hole may not necessarily do so when the plate is in place. Before finally pressing home the pins, try the run of both trains. If the run of all wheels is free, replace the gathering pallet, check that it is correctly positioned and pin it in place.

Replacement of the Motion Work

48 Reassemble the bridge and the motion work, fit the rack, rack spring, rack hook, lifting-piece, etc, and the date wheel if any, taking care to set the date wheel according to the mark you put on it before (see Figure 14).

There is no need to go into great detail about the date wheel, but one point is essential: it must not have surplus end play, and nor must the hour wheel, because otherwise the rack tail will foul it, and then the strike is disorganised. The rack tail must be free to fall on to the snail whatever the rotational position of the date and hour wheels.

More rarely found is a date wheel rotating anti-clockwise, geared direct to the hour wheel and with the same number of teeth. It may drive a moon disc by moving it twice a day, or it may operate a spring-loaded lever to turn a date disc mounted in the break-arch of a clock. Normally this is done by a pin or brass 'finger' mounted on the hour-wheel pipe, operating a series of levers under light spring, or perhaps gravity, to achieve the same result.

49 Now stand the clock on its seat-board in its correct upright position, and screw in the holding-down bolts, taking care that no part of the cables is trapped beneath the plates. *Do not* over-tighten these bolts or you will put undue strain on the pillars and may set up torque in the plates, so making the pivots tight and thus stopping the clock.

As stated elsewhere, these bolts should be firm enough only to hold the movement still during winding or other vibration.

Correct Setting of the Hands

50 The motion work has to be correctly adjusted, and as the minute-wheel pipe terminates in a small square on which the minute hand is mounted, the latter has a square hole in its boss. It can therefore be mounted in only one of four positions. Try each of these and select the best-fitting one.

The setting of the minute wheel and the motion wheel has already been described in detail; but when the dial is put on, you might find that the clock strikes either just before or just after the hour mark.

Final setting, as mentioned earlier, may then be obtained by gently bending the pin on the motion wheel towards or away from the lever on the lifting-piece, to allow for the drop to occur just as the minute hand reaches the hour mark on the chapter ring. Do not try to achieve this by bending the hands — only vandals do this — especially not the minute hand, which should be slender and straight (or if of the serpentine type, slender and evenly curved). The correct position of a rigidly mounted hour hand must be obtained by making the hour-wheel teeth engage the motion-wheel pinion in the correct position. This, of course, has to be done before the dial is put on.

Only where the mounting is frictional may the hour hand be positioned by holding the minute hand at the hour point, and gently moving the hour hand till it indicates an hour, ie points direct to one of the chapters. The snail will automatically set correct, but care must be taken not to put any undue strain on the teeth of the hour wheel when doing this.

Dial Re-fit

51 Having assembled the dial, cleaned it and re-silvered the rings, etc, fit it back into position on the clock. It is as well to slacken the seat-board bolts before putting back the dial plate, and to tighten them again afterwards. This allows proper positioning of the movement from front to back, so that there is no outward pressure on the dial plate. Pin the latter firmly in the manner described, remembering that the plate should not move in relation to the clock.

52 Now mount the hour hand and the minute hand, put on the hand collet and pin it in place. If the clock has a centre seconds hand, the minute hand is held not by a pin but by a slide washer fitting in a groove at the end of the cannon pipe. The motion work is also quite different.

Pulleys and Cables

53 The reassembly of the clock is now nearing completion, but we have not dealt with the pallets, pulleys, cables and weights.

You will see that to the left side of, and beneath, each barrel a rectangular hole is cut in the seat-board through which a cable passes. Having attached the cables to the barrels, thread them down through their respective holes, through the loop on the pulley and up again to a small hole in the seat-board to the right of each rectangular hole.

Thus the strike cable will emerge under the centre of the movement, and the going cable will come up to the right of the movement. If these small holes are not too large the cables may be knotted above them.

Cable Length

To adjust cable length, set the seat-board in a suitable vice. With the cable fully unwound, fit the clock key on the winding square and wind up the cable correctly in the grooves of the barrel by letting it run through the left hand, and exerting enough tension to position it correctly. Watch carefully till the barrel is 'full' with no overlap at all. If the pulley is still some way below the board, pull the cable at its knotted end until the pulley is close to the board, and then re-knot the cable. Take one or two loops in its end, make a final inward loop to form a tie-band to hold the others, and you have a good stop for the line. Alternatively, if you have fitted steel cables which are very springy, especially when new, each line may be cut to the right length and each end knotted over a short piece of ¼in (6mm) pegwood. If the lines are left too long they over-fill the barrels and may well foul the clicks or slip over a shallow barrel-end.

The seat-board holes for the ends of the cables should be roughly central in the breadth of the board, or slightly nearer the dial side rather than the back. Similarly the barrel-ends should for preference be at the dial end of the barrels; in general, however, they are not, as you may see. The reason is that the weights, as they descend during the eight days, should keep as clear of the pendulum as possible and also avoid fouling the lower edge of the trunk door of the clock case. It is a fault in the right direction to permit the weights just (and only just) to touch the door of the case, provided the door locks properly, especially if the clock is standing on a carpet with felt beneath it: this light touch hinders any tendency the weights may have to swing from side to side. If the case is not rigid, it will imperceptibly move with the swing of the pendulum, and the weights will also move; and five days or so after full winding, when they reach the same length as the pendulum they will cancel it out. So the clock will stop.

It may be accepted as a general rule that if the clock case moves the clock will stop; certainly it is unlikely to keep good time. The case is much more efficient if screwed to the wall. Also, think of the added security achieved thereby.

Positioning the Clock

54 It is therefore best to select a spot in the house where the grandfather clock can permanently reside. Any clock of some age will already have several holes in its back-board where previous owners have tried to fix it to old 'lath and plaster' walls. A place with an 'upright' within the wall to which the clock could be screwed is needed in such a case.

Assuming your wall is a solid one, mark it through one of the more central holes in the board. Move the case away, drill a hole at the mark, and plug it ready to receive a screw (brass, with a fairly large head).

If there is a skirting board along the base of the wall, obtain a flat batten of wood some 8 or 9in (220mm) long, about 2in (50mm) wide, and preferably a shade thicker than the skirting board. Drill a hole through the centre of the batten large enough to allow a 2 or 2½in (65mm) screw to pass through it right up to its head. This batten will hold the trunk of the clock clear of the wall by a little bit more than the thickness of the skirting board. Thus the clock will imperceptibly tilt forward enough to ensure that the pendulum swings clear of the back-board of the case.

Adjust the case against the wall so that the plugged hole is visible, tilt it forward, place the screw through the back-board, put the batten on the screw so that it will lie horizontally against the wall. Tilt the case back again, guiding the screw-point into the plugged hole. Turn the screw almost home, but not quite. Test the uprightness of the case with a builder's level, and if it is out of true, gently tap the side of the plinth by hand to move it to one side or the other until the level shows true. Then turn the screw home tight.

Test the level across the cheeks of the trunk where the seat-board rests. These cheeks should be rigid and level.

The Trunk Hood and Case

55 Try the position of the hood of the case, because any ill-fitting or looseness here should be rectified. See that the joints between the hood and the back-board are as closely fitting as possible, to exclude dust. Some owners have fitted their clocks with felt strips to prevent the entry of dust, which is not a bad idea provided that they are not visible.

Check the finials for firmness, and if the centre one is missing (they have often been removed in order to clear the ceiling) it has probably been put away safely. Find it, put it in a small polythene bag and, with a drawing pin or two, fix it inside the case; then it will always be where it should be — with the clock. Fix a small hook just inside the door on which to hang the key for safe keeping. Do not tie a string to the key because sooner or later it will catch on one of the hands of the clock during the winding, and damage it.

Finally, if there are any veneers or inlays on your clock which need attention, carefully glue them in place with one of the white resinous wood-glues readily obtainable, wiping off any surplus with a damp cloth while it is still wet.

Where mouldings on the hood, the door or the plinth have loosened because the original glue has perished, carefully examine the piece to see if a previous owner has tried to 'nail it back'. Panel pins or fine brads are often found, and of course iron or steel should never be used in oak because they will soon be rusted in. If you find rusted nails, their heads punched in and covered with stained putty or filler, the moulding must be freed and carefully levered up with a fine knife-blade. Gently lever up the piece — first at one point, then at the other — till it is free, taking care not to mark or damage the wood where it is exposed. The nails may break, which is not a bad thing. Generally, however, the nail heads can be drawn through the moulding on the inside so that no damage is seen from the front. When the piece is off, extract the nails from the case, or if they will not come out, cut them off flush and punch in the remaining ends just below the surface.

Now with a knife-blade or a piece of glass scrape off all the old glue from both the case and the moulding. Test that the moulding seats well and also fits any bevelled joint to another moulding. If all is as it should be, spread the glue carefully with a knife-blade on both surfaces, so that all parts are covered, and press the part in place, holding it there with suitable small cramps bearing on some sort of pad or pads to protect the wood.

Large pieces such as the plinth moulding may be held by weights while the clock case is lying on its back or side on the floor of the workshop. Wipe off any surplus glue as before.

56 It is an even bet the clock originally stood some 7ft (210cm) or even 8ft (240cm) tall, and that the plinth will have been altered to shorten it. Again, if the clock is what is known as a 'farmhouse' clock, its case made of good solid English oak, it may have stood for years upon a flagstone floor and been weekly swilled with water as the farm wenches washed the floor. Its feet will have rotted long ago and been mended by a local carpenter or cabinetmaker. Work by the latter will most likely be excellent and very hard to detect, but less careful alteration, with little attention to choice of wood and grain, will be easy to see. The practical man at the time — and after all the clock was then worth only a few pounds — repaired the case with a view to everyday efficiency rather than to any aesthetic considerations. If you feel that the plinth of your clock shows glaringly misconceived repairs, take it to a qualified craftsman cabinetmaker, and give him full detailed instructions in writing as to how you want the plinth rebuilt. First go to your local library to study pictures and photographs of contemporary clocks in horological books. Probably one of the best is *English Domestic Clocks* by Cescinsky and Webster, which not only gives types and designs but also approximate dates.

It is often easy to strengthen a longcase by gluing battens or small blocks of wood inside it. Where the lower end of the back-board, or perhaps the flank, of a case has split due to the effects of central heating, you can repair it by gluing and screwing a batten inside the case, well clear of the weights and pendulum.

First position the case so that you can get at it easily. Obtain a carpenter's adjustable long cramp, and with it gently squeeze the wood till the crack has closed as much as it will. Take care to protect the case by inserting bits of hardboard or thin plywood under the 'grips' of the cramp. Leave the cramp in place while you complete the following. With a rule, measure the place where you want to fix your batten, which should be dry and seasoned — preferably oak, but a good hard pitch-pine will do. Keep it by the central heating for a week or so before use, but *do not* specially heat it. The object of this, of course, is to get the woods of clock case and batten of more or less equal moisture content.

Cut the batten to size, place it across the crack, mark the places chosen in sound wood for the screws, remove the batten and where the marks are, drill holes large enough to clear the screws. Counter-sink the holes.

Carefully measure the thickness of the back-board, or flank as the case may be, and also the thickness of the batten. Press a brass screw home in a counter-sunk hole, seeing that it does not protrude from the batten more than the thickness of the back-board, etc. When screwed home none of the screws must penetrate through the board into which it is screwed. If they might do so, find shorter ones.

Now glue the batten, spreading the glue evenly, and screw it down, with the long cramp still in place of course. After twenty-four hours release the cramp.

The Recoil Escapement

57 Once again we must return to the movement, this time the escapement.

Most grandfather clocks have a recoil or anchor escapement. This consists, firstly, of a wheel of (generally) 30 teeth, each tooth being radially straight on one side, and concavely curved on the other; and secondly, of two pallets mounted on an arbor. An anchor-shaped piece of steel mounted on, and at right angles to, an arbor is activated by the teeth of the wheel to give enough impulse to maintain the momentum of the swinging pendulum. Here again the mathematics of the matter are a study in themselves.

58 As the arbor carrying the anchor or pallets extends through a hole or gap in the back-plate, its rear-end pivot is mounted in the back-cock, of which more later.

The clock ticks once per second and the wheel-teeth are released one at a time as the pendulum swings. This is so simple that it is accepted by everyone, even youngsters, but few bother to think about what is really a marvellous invention, certainly for its day in the seventeenth century.

If you set a freely hanging pendulum swinging by itself, it will do so for a long period before it comes to rest. If a well-made heavy pendulum of 14 or 15lb (about 6kg), properly and rigidly hung, is set going after breakfast, it will still be perceptibly going after dinner, some eleven or twelve hours later. This indicates that the loss of momentum is *very* slow. We will not go into the reasons, but if we could replenish this loss we could perhaps achieve something like perpetual motion. If the pendulum we have mentioned above was given a very small push every hour or so, it would indeed go on swinging. But instead of pushing it like that, suppose we had some device to give it the very smallest of pushes at every swing, first one way then the other; then we could establish a sort of balance between 'push' or impulse and momentum loss. If we had an ideal clock completely free of all friction, we could make the impulse equal to the loss of momentum; but this cannot be.

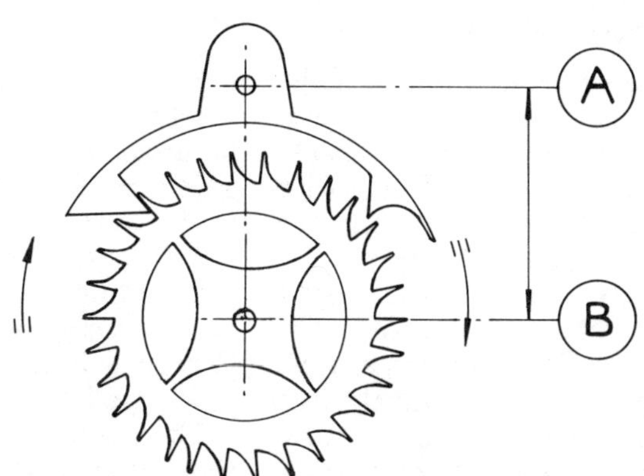

A CONVENIENT DISTANCE FROM A TO B IS THE WHEEL RADIUS MULTIPLIED BY 1·4. THE SPAN OF THE PALLETS SHOULD BE 7½ TEETH (THE DRAWING SHOWS 8½, WHICH WILL WORK QUITE WELL). THE PADS OF PALLETS AT THEIR TIPS ARE AT A TANGENT TO A CIRCLE CENTRED AT A. (IF THE ARBOR A IS TOO FAR FROM B, THE PENDULUM WILL NOT SWING OUT WELL.)

SECONDS RECOIL ESCAPEMENT (SECTION 57)

Figure 21

We drive the clock by a weight to supply the impulse, but we have to make the weight heavy enough to overcome all friction; and as this friction is a variable factor we allow a margin of extra weight. Thus, the impulse force to the pendulum is a bit more than the loss of swing, and the excess swing gently pushes the escape wheel backwards very slightly at each oscillation. Hence the name 'recoil' escapement.

Entrance and Exit Pallets

59 From the front of the clock look closer at the pallets, with the escape wheel rotating clockwise. The left-hand teeth are coming up, while the right-hand ones are going down. The left-hand pallet surface is roughly horizontal and the right-hand one almost vertical. They are called the entrance pallet and the exit pallet respectively, and the angle of their surfaces or 'pads' is important. It is the tip of the curved side of each tooth which falls on the pallet, the surface of which (at least in part) lies at a tangent to a circle centred on the pallet arbor; the circle has a radius designed to suit the escape wheel. Most pallet arbors are set at a distance of 1.4 times the escape-wheel radius from the escape-wheel centre.

Action of the Escape Wheel

60 In order to see the action, mount the escape wheel in the plates by itself, and mount also the pallet arbor and back-cock. Gently rotate the wheel clockwise with the left hand, while holding the crutch with the right, allowing it to oscillate slowly. If you let go the right hand while continuing the rotation, the pallets will oscillate the arbor and crutch.

Now fix the pinned plates lightly in a vice, or in some way that holds them steady to allow you the use of both hands. Hold the crutch in one hand and apply light rotational pressure with the other to the escape arbor, at the same time moving the crutch slowly from side to side.

As you are facing the clock, move the crutch to your right. You will see that one of the teeth of the escape wheel contacts the entrance pallet. If you continue to 'swivel' the crutch to the right, the entrance pallet will push the wheel back a bit until the pallet point can go no further into the wheel.

Now very slowly move the crutch to the left and watch the tooth on the face of the entrance pallet. It slides towards the tip of the pallet, while the latter retreats on a circular path till the tooth is cleared. The tooth is said to 'drop off'. Just before the tooth drops, when it is on the very tip of the pallet, hold it steady, and observe the position of that other tooth which is about to contact or 'fall' on the exit pallet.

If there is a very loud click or tick, the amount of fall may be excessive. Continue to observe the exit tooth and you will see that the same thing happens here too. The tooth will at first slide up the face of the pallet, thus turning the escape wheel slightly backwards.

When the crutch is moved to the right again, watch the tooth carefully till it reaches the tip of the pallet, and as it clears the tip, note if there is much fall on the entrance pallet by the next entering tooth.

Pallets and the Depth Tool

61 The following points will be found useful to the handyman in making adjustments to the pallets.

Firstly, the use of the depth tool should be understood, for the student will then gain some knowledge of the exact requirements (see also Sections 34 and 37).

The clockmaker's depthing tool has four sliding parallel rods or runners, each of which may be fixed by a thumb screw. Within a hinged frame two rods are in line, parallel to the other two, also in line.

The frame is hinged and sprung, and the distance between the two sets of rods may be varied by a fifth thumb screw. Each runner has a true centre point on the outside, a male centre, and a centre recess or female centre on the inside end. Thus by sliding the rods apart, a wheel arbor may be mounted in each set of runners, and a wheel in any one set may be made to

Movement seen from the going-train side. Note the brass extension on the date wheel, which moves the date ring and is shaped to clear the snail and hour wheel. Note also the lifting pin on the motion wheel (Section 19)

Front view of an 8-day movement. The rack is in the 12 o'clock position, its arm cut away to clear the winding square. Note the marks where the rack tail has ridden over the surface of the snail (Section 20)

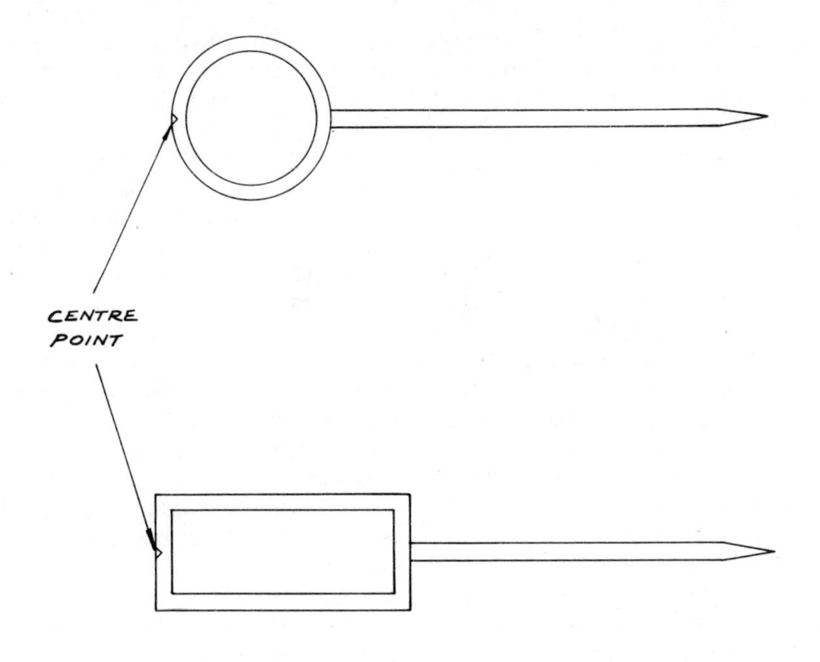

LANTERN RUNNERS FOR A DEPTH-TOOL.
THE OVERALL LENGTH IS SHORTER THAN THAT
OF A NORMAL RUNNER, BECAUSE LANTERN RUNNERS
MUST BE FITTED FROM THE INSIDE OF THE DEPTHER

DEPTHING TOOL (SECTION 61)

Figure 22

engage with a pinion in the other. Both arbors can be removed without disturbing the setting of the runners.

It is obvious, however, that if you tried to put a pallet arbor, complete with pallets and crutch, into such a tool you could not oscillate it because the crutch would come up against one side or other of the opposite runner. If that runner was divided, and shaped as in Figure 22, then the escape wheel could be mounted in such a way as to permit the escapement to operate as it would in a clock.

Mount the pallet arbor in the depth tool, in the far-side runners, with its crutch through the gap in the nearside runner; and mount the escape wheel in the nearside runners, just as they would be when mounted in the clock. Slowly bring the arms of the tool together, holding it so that the crutch hangs down. Rotate the wheel till the pallets operate as they should, and when the action is as close as possible, and all the teeth of the wheel clear the pallets as closely as possible without fouling any of them, carefully remove the parts from the tool without disturbing the setting. The thumb screw which controls the distance between the runners must *not* be touched while the parts are being taken out of the tool.

The tool may now be used to mark the

plate to show the correct distance between the two arbors. It is no use, however, carrying out this operation with an escape wheel which is not true; so next we look at escape-wheel problems.

Escape-Wheel Repairs

62 You may see that some of the escape-wheel teeth are worn and that their lengths vary; and if this is true of even one tooth, it is no use proceeding further with the escapement until it has been corrected. All the teeth must be true, equidistant and equal in length. Escape wheels in general should be as light as possible, and this entails a certain frailty, so they should be treated with care.

The wheel we are dealing with has 30 teeth, so that the distance in degrees between each tooth is 360 divided by 30, namely 12°. Obtain a piece of white laminated plastic sheet and, with a large protractor and a very fine ballpoint pen, mark a centre point. From this point draw five or six radial lines, each *exactly* 12° apart, and each extended through the centre point in the opposite radial direction. You could divide the whole circle into 12° segments but this is not necessary. The use of a large protractor enables you to make wide marks exactly on each 12° point. Drawing these lines through the centre out to the other side is likely to give more accuracy, which is very important here. Now drill a hole at the centre point of the size of the escape-wheel arbor. If the wheel is mounted on a collet, as is virtually certain, the hole will need to be the size of the small side of the collet, the object being to permit the wheel to lie as close to the surface of the laminate as possible.

Place the escape-wheel arbor in the hole, and rotate it till the radial or straight sides of the teeth coincide with the lines you have drawn. Suppose you have six lines, and on lines 1 to 3 the teeth coincide exactly, but the teeth on lines 4 and 6 are slightly bent. These can be carefully adjusted with small flat-nosed pliers till they do coincide with the lines. See if the teeth on the opposite ends of the lines also coincide, and rotate the wheel in 12° changes several times, till you are satisfied that all the teeth are angularly true and that each radial side is accurate.

You still have to check the length of each tooth. Remove the escape wheel and mount it again in the depthing tool, this time alone, with the runners wider than the wheel. See that the wheel is free to rotate but has no side play at all. Now close the opposite runners (in line) and reduce the distance using the fifth thumb screw, till one tooth of the wheel can only just touch the opposite runner, with as light a touch as possible, ie the minimum you can adjust; mark this tooth lightly and clearly.

Now slowly rotate the wheel, trying each tooth in turn, because ideally each of the other 29 should have the same degree of 'touch' as the one you have tried. If any of the teeth touch tighter or closer than the one you have set, the tip of the tooth concerned may be gently filed while the wheel is still in the depthing tool. Constant trial will soon ensure that all the teeth have the same clearance. Clean away all filings.

You will soon see by experience that other factors emerge, eg the tips of some teeth are thicker than others. And, provided the arbor and pivots are not bent, if teeth numbers 2 to 14 touch the runner readily, whereas numbers 16 to 29 do not touch at all, then the wheel is not mounted true on its arbor, and you will have to true it in a lathe and re-file the teeth to shape, filing only the curved side. While the wheel is still in the lathe, adjust a graver tool on the face of the wheel to the base of the deepest cut between the teeth, and scribe a faint circular line right round the wheel. The bottom or depth of each tooth should be filed to this line, but do not file the radial side of any tooth.

If you find that the arbor itself is bent, you will have to obtain a new one. Again, if after adjusting the length of the teeth any are found to be too thick at the top, then the curved side must be gently filed to restore the correct shape and the fineness of the tip. Special files may be bought for this. After filing, ensure that the curved surface is polished smooth.

63 Now you have a true escape wheel with all its teeth the correct distance apart, and all exactly the same length.

Reverting to where we left off, you can now test the escapement properly. Set the arbors in the depthing tool again, adjust the depth and the fall of the teeth, and reduce the clearance to the point where the escapement does not work, that is to say the wheel will not rotate.

Widen the gap in the tool till the pallets just clear the teeth, and till the whole wheel turns freely, clearing the pallets with the least possible gap. If the pallet, particularly the exit pallet, descends on the tip of any tooth, that tip is probably still too thick and will have to be filed a bit more. Use a very fine file, and leave each curved surface clean and smooth.

If the fall on the exit pallet is too great, the pallets need closing slightly, or else the exit pallet needs building-up. If, when the parts are in the clock, the fall on the entrance pallet is too great, the pallet arbor must be brought a bit closer to the escape wheel.

Remember that the pallets are hard and brittle; they break very easily, so never attempt to bend or adjust them without first annealing them. Once softened, place the pallet anchor, concave side up, in the gap of a vice fitted with copper or wooden protection jaws, each jaw being under the end of one of the pallet arms. The vice should be wide enough to support only the extreme ends of the pallets.

Using a brass punch give a light tap on the inside centre of the anchor, and try the fall again. Repeat this procedure till the fall is correct.

Where either pallet needs building up, cut a piece of annealed spring steel to the shape of the pad of the pallet in breadth, and about half an inch longer. This extra length is for ease in manipulation when positioning the piece, and to ensure that the soldered joint is as thin as possible — the surplus end of the piece being held with suitable pliers, by means of which pressure can be applied while the piece is under heat. Carefully clean both the shaped end and the pallet pad, heat them, and silver-solder them together as closely as possible. When cold, cut off the surplus spring, file the sides of the built-up pallet to shape, polish and finally burnish the acting surface. (The latter should not be filed.)

This is quite a common way of effecting a repair but it is really only a botch, because the pallets cannot be properly hardened again. The proper way is to make new pallets, but that is rather beyond the scope of this book. A repair such as that described should last for a few years if it has been well done.

Note: when heating the pallets care should be taken not to heat anything else. Pallets are usually silver-soldered or rivetted on to their collets, the latter being soft-soldered on to the pallet arbor. This allows some degree of change of lateral position of the pallets to bring a new surface into contact with the wheel-teeth. The anchor may be held near its middle in a large pair of pliers, the grip of which will absorb heat and help to concentrate it on the pad while protecting the centre joint.

Replace the parts in the depthing tool to check the action, and if this is correct, remove them. Now with the points of the depther test the distance on the *front* plate between the escape-wheel pivot hole and the pallet-arbor pivot hole. The points of the runners should fit exactly into these two holes, absolutely centrally. If they do not, then the pallet-arbor pivot hole will have to be adjusted accordingly, as described in Section 35.

Let us assume, however, that the holes are correctly depthed, and proceed.

Back-Cock Adjustment

64 The rear end of the pallet arbor is mounted in a pivot hole in the back-cock, the latter being screwed and steadied (by steady-pins) to the back-plate. Thus there is a final means of adjustment here by moving the back-cock up or down, albeit infinitesimally.

Try the escapement with the back-cock as it now is, having tested with the depther the distance between the two pivot holes. This is tricky, as the hole in the plate is on a different plane from that in the back-cock, and it is essential to hold the tool upright.

Ordinarily, when a plate is clean and shiny, you can put the depther upright by aligning it with its own reflection in the plate. Here, however, the plate can be clamped to the edge of the bench, columns upwards, with the back-cock clear of the edge of the bench. It is then fairly easy to set one runner point in the escape pivot hole, and lengthen the other runner till it reaches the pivot hole on the inside of the back-cock. Any irregularity will then at once be apparent (see page 71).

65 The back-cock should be moved only a very little, and it can often be done, when the screws are loosened slightly, by tapping it with a brass punch. Very slightly loosen the two holding screws — not as much as 'finger' loose — and tap the part up or down to obtain a good escape action with free impulse. Tighten the screws. This action merely moves the steady-pins a bit, and may make just that final difference.

Some jobbers remove the pins, file great ovals in the screw holes and make the back-cock adjustable, so to say. This is bad. The alignment of the pivot holes is affected and also the steadiness has been lost. The weight of the pendulum, as a result of the vibrations of the clock hammer, after a while may move the back-cock and upset its adjustment; a well-fitted one, with good steady-pins, is the only safe bet.

Incidentally, the fitting of the steady-pins is worth mentioning. When the proper setting of the pallets has been obtained, make sure that the back-cock is screwed on tight. Separate the plates carefully, leaving the back-cock in place; then take out the escape wheel and the pallets.

Stand the back-plate on its columns, ie with the back-cock upwards, and drill two *small* holes through both back-cock and plate, one on each side. The old holes (if any) will not coincide, but maybe those in the plate *or* those in the back-cock can be used as guides. Any old holes not used should be plugged.

Take care to drill the new holes at right angles to the plane of the plate to which the part is fastened. Separate the two parts, and in the one to be pinned (not the plate) drive in two pins to a tight fit. Slightly counter-sink the plate, and try the fit. The driving-in of the pins in the one part will have made a slight burr on the side of contact, and the counter-sinking will clear the burrs to permit a close fit.

Cut off the pins to size, file the ends, and clean up the outer ends flush with the pinned part. If the pins fit too tightly in the plate, broach out the plate holes till the fit is perfect.

Pulleys

66 An average pulley consists of a 1½in (38mm) brass wheel, solid — ie without crossings — mounted on a short steel axle, from the ends of which hangs a loop of brass or iron wire. Loosely suspended from the loop is a hook, on which the weight hangs. It is preferable to mount the hook on the loop of the pulley rather than on the weight itself: it is easier to hang weights on a dependant hook than to fiddle a loose hook on to the pulley loop. It is much easier for the hook to face to the right (have its opening to the right) as you will have the weight in the right hand — unless you are left-handed.

If the axle of the pulley is much worn it should be replaced. The ends of the loop, fastened to the axle, will be rivetted on to a shoulder in the end of the axle. On no account, therefore, try to punch out the axle. File the end to remove any rivet effect and lever that end of the loop off the end of the axle.

Bend the loop, under heat if necessary, to allow the pulley to be removed, then file the other side, and remove the axle from the loop.

Examine the hole in the centre of the pulley, and if it is not true, drill it out only sufficiently to obtain a true cylindrical hole at right angles to the plane of the wheel. Fit a new axle to this, cut it to the correct length and, in a lathe, turn shoulders at each end to make a tight fit in the ends of the loop. These ends should still be good and undamaged.

Now make a shallow counter-sink on the *outside* of each hole in the loop ends, assemble the pulley on its axle, fit the loop ends and rivet them.

The slight spread caused by rivetting will allow the burr to fill the shallow counter-sink, and the loop will be held firm. Tidy up the job with a fine file and polish all.

Alternatively, you may go to a supplier of horological materials and buy a beautifully made (by machine) pulley complete, and it will last till after your great-grandchild is dead. You will, however, have lost something by doing so — apart, I mean, from the £2 or £3 it will have cost you.

After you have fixed a hook on the pulley loop (brass S-shaped ones look nice), thread the cables as described above, and see that each cable lies cleanly in the groove of its pulley before hanging the weight.

Weights (Poundage)

Weights for longcase clocks vary, but those for the average 8-day grandfather are usually from 12 to 14lb (about 5 to 6kg). In many clocks the two weights are the same; in others there is 2 or 3lb (1kg) difference. Where they are different, mount the heavier weight on the strike-train pulley; if you do not, the strike will sound annoyingly slow. See what anyone else in the house says if they happen to be telephoning at noon.

The weights for a month, 3-month or 12-month clock are correspondingly much heavier, but we are not dealing with these here. Conversely, precision clocks have much lighter weights.

For preference, weights should be dome-topped, because these collect less dust. The hook or staple loop by which they are suspended should be central, so that each weight hangs vertically. They are usually of lead or iron, the former being easier to cast. Brass-cased lead weights are found on the better-quality clocks. Iron weights often look unsightly, with rust marks and blotches, but a coat of ordinary school blackboard paint improves their appearance.

Fitting a Suspension Spring

67 Look at the pendulum to see if the suspension spring needs attention, that the bob is free to slide and that the rating nut is readily adjustable. The suspension spring should be true, flat and show no sign of twist or damage. If it is in any way imperfect, you will have to get a new one from a supplier, because it is no use trying to make a new one out of an old bit of spring. It can be done, of course, but the purchased one is already made up — with brass ends, some with one end clear and others with a brass block ready tapped for the pendulum rod. The former have the advantage that they may be cut to the exact length of the old one, and are reasonably easy to fit to the existing rod.

Remove the rating nut and also the bob from the rod. Put the latter — assumed here to be a wire one — firmly in a vice and unscrew the brass block to which the suspension spring is fastened. The spring is fixed so that it can hinge easily from back to front, but can only bend from side to side: this is an important point.

Gently file one end of the pin holding the spring in the brass block, and punch out the pin to free the spring. The pin is sometimes of brass and sometimes of steel; if it has been badly put in, you may have to drill it out, carefully so as not to damage the hole.

If the brass block is badly worn, too thin for the crutch, or out of true, then it may be best to fit a new brass block altogether. Obtain a suitable piece of brass, and first see that the slot in the crutch is truly squared — that its sides are parallel. Try the fit of the brass block in the crutch. If the old block is loose, discard it. A new one can be filed flat and true on both sides to form a free fit.

A good test is to stand the brass block upright on a flat surface and lower the crutch over it, raising it up and down to see that there is clearance but little or no side play. It becomes evident that each side must be carefully filed so that its surface is flat, to contact the side of the crutch all along its length. This will aid time-keeping by preventing the pendulum 'wobbling' when swinging.

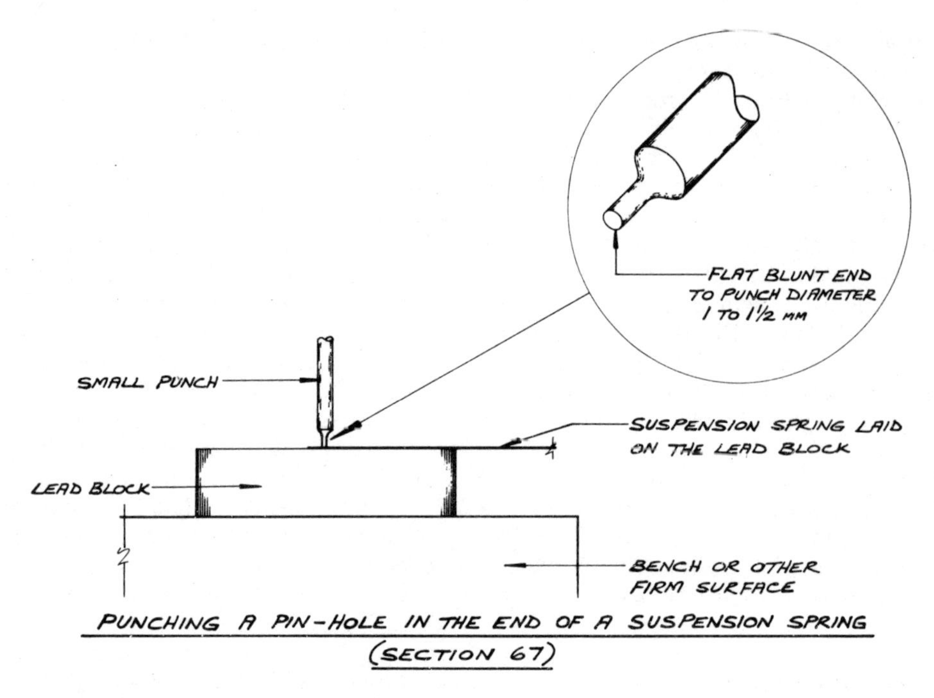

Figure 23

Now drill and tap one end of the brass block, and fit it to the pendulum rod. Make a good tight fit with the block positioned at right angles to the plane of the steel block at the other end of the rod.

Slot the opposite end with a very fine slotting saw (a piercing saw is not suitable unless you can cut dead straight) and drill a small hole for the fixing-pin. Place the suspension spring, cut to the length of the old one (and checked and measured by the length of the crutch in the back-cock), in the slot in the brass rod, and with a fine scribe mark the position of the hole to be made in the spring. Remove the spring, place it on a block of lead – a clock-weight perhaps – and, with a fine steel punch pressed firmly on the marked spot on the spring, punch a small hole. The punch must not be pointed; it must have a clean, flat, circular end, and it should be struck with just enough force to penetrate the lead. A nice clean round hole will result (see Figure 23).

Slightly counter-sink the outside of the holes in the brass block to take the rivetted head of the pin. Assemble the spring, put in a new pin, rivet it in place and tidy up the rivets flush with the block on both sides. Leaving the rivet heads proud of the block will cause a lot of bother when fitting the pendulum to the clock in its case. They will not pass through the pendulum crutch, because you have made the brass block to fit the latter. Take care (as always) when rivetting not to close the brass on to the spring so that it grips the latter. Remember the spring should be free to hinge in its own plane, that is from back to front. Finally, check that the block is firm and at right angles to the plane of the pendulum bob.

Pendulum Repairs

68 Now look at the other end of the pendulum rod and test the fit of the bob on the squared end-piece. The latter is screwed to the rod in the same way as the brass block at the top end. Into the lower end of the end-piece is screwed a threaded steel rod 3 or 4in (90mm) long, and on this is screwed the rating nut. The bob rests on the rating nut in such a way that it may move freely up and down according to whether the nut is screwed up or down, but cannot turn or twist. Many end-pieces are slightly tapered to aid this steadying effect. The bob itself is lenticular in shape and is made, in the best instances, of two concave brass discs brazed together and filled with lead. There is a squared hole from top to bottom, through which passes the end-piece of the rod.

Some bobs are of lead, faced with brass on the front; others are cast iron and decorated with painted designs. A highly polished brass pendulum bob swinging freely in a clock is a pleasant sight.

Before reassembling the pendulum make sure that the wire rod is straight, without bends or curves. It may be straightened by being hammered gently on a flat steel surface till all irregularities are removed. The rod looks better if painted with quick-drying shiny black lacquer.

Rating-Nut Mounting

69 Clock pendulums are often damaged by being dropped. The rating nut can lose its thread in this way, or its mounting can become so bent as to be useless.

Often the end-piece has been shaped to include the threaded portion, all in one piece; where the latter has been damaged it will have to be cut off and a new one fitted. First carefully remove the threaded part with a hack-saw, and file the lower end of the end-piece true and square. Then mark the centre, drill a small hole to take, say, a No 8 BA tap, to a depth of a little over ¼in (7mm), and finally thread it.

Select a piece of thin rod, just too big to enter the hole you have drilled, and put a thread on one end for about ¼in (7mm). An 8 BA fits most end-pieces, but their thickness does sometimes vary. As a No 8 is small and weak, the guiding factor is to put on as strong a threaded rod as possible — the pendulum will be dropped again before its life is finished.

If the new joint is too weak, you can always silver-solder it to add a little strength.

Screw the threaded rod tight into the end-piece, lay it on its side on a flat steel surface and lightly hammer each side to ensure that the rod is rigidly secure. Put the piece in a vice, rod upwards, and put a thread on the rod for most of its length, the same thread as that of the rating nut.

The thread used for the rating nut will be different from that used for fixing the little rod into the end-piece. You can always make a new rating nut, of course, to fit the thread you have put on the rod; if you do, make a fairly large one, hexagonal in shape. This helps with accurate rating, especially if you file nicks in every other side, one nick side 1, two in side 3, and three in side 5 (see Figure 24).

Assemble all the parts, so that the pendulum is ready for hanging. As we have seen, it hangs by the spring from the back-cock, the cheeks of which should hold the spring firmly without side play or movement of the brass cap. The brass cap should be made to fit into a suitable slot or depression cut in the top surface of the back-cock extension. This ensures that it cannot easily slip out of place and let the pendulum fall down.

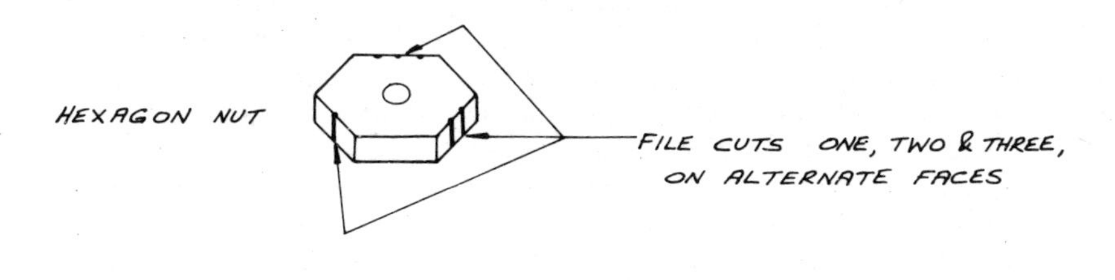

Figure 24

Pendulum Length

On the underside of the back-cock note the point where the spring bends as the pendulum swings. (You do not need to be so exact, however; the underside line of the cock will do.) The distance from this point to roughly the centre of the pendulum bob should be 39.14in (approximately 1m). There are all sorts of mathematical things to say about this (and also about circular error), but they do not concern us here.

This measurement varies according to the latitude of the earth, but 39.14 inches or 1 metre will give us a good starting-point from which to rate our clock, ie a seconds pendulum clock. The whole rod to the end of the threaded extension will be about 41 or 42in (1,070mm) long.

Putting the Pendulum in Beat

70 We assume now that all repairs are complete, and the movement is mounted in the case ready to receive first the pendulum, then the weights. It is wise to have the cables fully unwound so that there is no danger they will get loosened, crossed over or off their respective barrels. Fit the pendulum in place, seeing that it is properly seated in its little groove. Now wind on the cables correctly in their barrel-grooves, put on the weights with the cables running rightly in their pulleys, and swing the pendulum until the escape wheel begins to rotate.

After a little while to allow the pendulum to 'settle', note whether the beat or tick is even. If it is uneven, the pendulum is said to be 'out of beat', and it should be put 'in beat' by very slightly bending the crutch to one side or the other until the beat is perfectly regular. You have already made all the teeth of the escape wheel true, but it would be a miracle if all thirty teeth were identical. The human ear will soon detect that all the 'ticks' are not the same either, but this need not affect the rhythm. You can readily put a clock in beat by listening to it and adjusting the crutch, which should be bent away from the pallet which is most deeply engaged. In other words, if the clock is put in beat by being lifted on one side, then the crutch should be bent towards that side. Often the very slightest touch will suffice, and the operation may be done with the fingers of both hands from each side of the clock when facing it. The adjustment should be done with the clock in situ, in its case, especially if the case stands on an unlevel floor. Take care to bend only that part of the crutch which is vertical.

The horizontal slotted foot of the crutch should always be at right angles to the movement. On an unlevel floor, small wooden wedges can be pushed under the plinth of the case to achieve steadiness, especially as some owners object to having their walls plugged as suggested in Section 54.

It is essential that the case of the clock should not move, and rigidity can generally be achieved by wedging the front so as to press the back of the trunk against the wall, or against the batten between the case and the wall. After placing any wedges, mark where they protrude outside the case, remove them, cut them back to the mark, leaving it visible — and then replace them under the case. This should prevent them being accidentally dislodged. Remember that if your house is burgled a clock firmly screwed to the wall is much more likely to be left there by the intruder!

Rating the Clock

71 Dial the speaking-clock for the exact time and set the clock going. At the same time next day note down how much it has gained or lost. If the clock has gained, the pendulum is too short, and the rating nut must be lowered by a noted number of turns. Check again at the same time the following day, and so on till the clock keeps fairly good time — the loss or gain being related to the number of turns or half-turns of the nut. This is where the hexagonal marked nut comes in useful, but to avoid error write down all your primary adjustments.

Now do the rating once per week, again noting the relationship between the variation and the number of turns of the nut. After a week or two the time-keeping will be almost perfect, affected only by temperature, and if you set the clock to lose a few seconds per week the error (for domestic purposes) may be rectified when you wind the clock.

Domestic Care

72 The case should be polished with a wax polish once in a while, but *never* clean the dial with metal polish. The dial is lacquered and should need no attention for about half a century. A light dust over with a soft hairless duster once a month is enough. As the hood door is opened once per week to wind the clock, dust is bound to get in, especially if the clock stands near a door or window. Vacuum cleaners often emit dust which is so fine that it will settle and stay on a vertical surface. When it falls on wheels and pinions it is sometimes dislodged by their motion, but only if the parts are dry and free from oil. This is why pinions and wheels should never be oiled – only the pivots, and then sparingly with clock oil only.

The wheel and pinion motion in a clock is so slow that wear without oil is less than wear with oil – quite the opposite of, say, the gearbox of a motor car.

THE 30-HOUR CLOCK

73 All the foregoing remarks apply to the average grandfather clock, but only one type, albeit the most popular one. Another is the 30-hour endless-chain type which is generally wound up by its owner just before he or she goes to bed each night.

It will be obvious to the mathematically minded person that the gearing for a 30-hour clock must differ from that of an 8-day clock. The first thing you notice is that the dial of such a clock has no winding holes, and more often than not no seconds-dial. The second thing is that the movement has no barrels.

In some of the older clocks there are no front or back plates as such, but 'ceiling' and 'floor' plates, to coin a descriptive phrase, joined at the corners by pillars of iron and square in cross-section. This is known as a 'birdcage' movement, and its trains are set one behind the other.

The dial may show no minute marks but will show quarter-hour divisions on the inside rim of the chapter ring. Some clocks have only one hand; it has an elongated tail for easier turning to re-set the time. On the back-plate of the more normal 30-hour clocks – sometimes elsewhere – will be found a disc with steps cut in its periphery, the distance between the steps varying. This is called a 'count wheel' or 'locking plate', and it regulates the number of strokes made by the bell-hammer.

74 These 30-hour clocks normally have only one weight, and the removal of this and the pendulum will leave the movement free to be lifted out on its seat-board. The chain (sometimes a rope) is 'endless', in the form of a large loop, and it passes over two sprockets in the clock.

Let us follow its course, starting at the pulley to which the weight is hooked. Upwards from the pulley over the left-hand sprocket (left as you face the dial) in a clockwise direction, down through the seat-board for some distance to a small ring counter-weight, through this and up again to the right-hand sprocket, over this anti-clockwise and down again to the pulley, where it links up with its other end, so to speak.

The sprockets and the pulleys vary according to whether a rope or a chain is used. Clock ropes are specially woven for the purpose and are less liable to fray than are other types. To splice them requires skill, but modern adhesives sometimes get over the problem (see Section 83). The pulleys for ropes have rounded grooves, as do the sprockets, the latter being fitted with short spikes to prevent the rope slipping. The small counter-weight keeps the rope in tension so that the spikes may grip more readily.

For chains, the grooves in both pulleys and sprockets are shaped to fit the links. Each link in a chain is at right angles to the next one, and accordingly the groove is in two shapes: the deeper and narrower ones accept the 'end-on' links, and the oval shallower ones receive the flat links; into each of the latter will pass a spike of the sprocket. You will see at once that the distance from the middle of one flat link to the middle of the next-but-one must be equal to the distance between two spikes. In other words you cannot use any old chain; it must be one to fit the sprocket.

Odd worn or stretched links can be removed, and others can be reshaped, but if the chain is much worn the safest way is to buy a new one. Take the sprocket to a materials dealer and get him to fit a chain of the right size. They are made at so many links per foot (305mm).

The right-hand sprocket is free to rotate on its axle in a clockwise direction, but is prevented from turning freely the other way by a 'click', of which more anon. The left-hand sprocket is rigidly mounted on its arbor and only rotates with that arbor. Thus, the chain hangs from the sprockets in two loops, one carrying the pulley and weight, and the other the small counter-weight – generally a ring of lead weighing a few ounces.

The main weight will in this case be from 8 to 12lb (about 3 to 5kg), according to what power is needed. Too heavy a weight may cause the pendulum to hit the sides of the case – very undesirable.

Pulling down on the extreme right-hand chain raises the weight, ie winds the clock, without affecting the pull on the left-hand sprocket, which is the one that drives the going train. Thus the timekeeping is not affected because the clock is not deprived of power while it is being wound. This is referred to in clockmakers' parlance as 'maintaining power'.

On regulator and astronomical clocks, maintaining power is essential; you find it on better-class 8-day clocks, but seldom on the ordinary 8-day grandfather. There are several forms, most of them relying on some kind of light spring coming into force, acting on the great wheel of the going train while the clock is being wound.

Removing the Chain

75 In dismantling the movement it will be convenient to remove the chain, although it is not essential to do so. The links are not welded so that one of them can be opened to allow the chain to be split.

Do this and remove the chain from the sprockets, observing at the same time the formation of the inside of each sprocket. The wire hook you made before, for removing the lines from the 8-day clock barrels, will be found very useful for removing the chain. The sprocket is 'wavy' as described above, and the spikes pick up the links of the chain to prevent any slip.

Now you are ready to dismantle the clock, and the first things to remove are the hands (or hand), and also the dial, as previously described (see Sections 10 and 13).

In passing it may be mentioned that where the clock has only one hand, the front bearing of the wheel carrying the hand is actually in the dial plate.

30-Hour Striking Details

76 Having done this, study the strike mechanism carefully. Note the degree of play between the lifting-piece and the upper lever it operates. The star wheel or the motion wheel, depending on whether the clock has one or two hands, contacts a lever mounted on an arbor across the plates. On a one-handed clock the lifting-piece is often operated by a twelve-point star wheel fixed rigidly to the hour-hand arbor and hour wheel.

The star is fixed to the side of the hour wheel itself, the bearing of the latter being, as stated, in the dial plate. The lifting-piece is also a detent, as it has an arm between the plates which performs two functions: to lift a further lever, and to catch a pin on the warning wheel. This further lever is mounted on a second arbor lying just above the first one, and protruding its squared end through the back-plate. It has a second lever with a detent at its end, and the function of this is to stop the hoop wheel rotating after the clock has struck.

On the outside squared end, outside the back-plate, is mounted another lever, its end bent downwards at right angles. This bent bit is about half an inch long, and is set to fall into the slots of the count wheel or locking plate. When the lifting-piece is raised, it in turn raises all the levers above it, because they are in one piece pivoted about the same centre. We shall see the reason for this a little later. The second arbor thus has three projecting arms (see Figure 25):

a) the short arm, which is contacted by the lifting-piece below it;

b) the detent, which arrests the hoop wheel;

c) the outside detent which runs on the

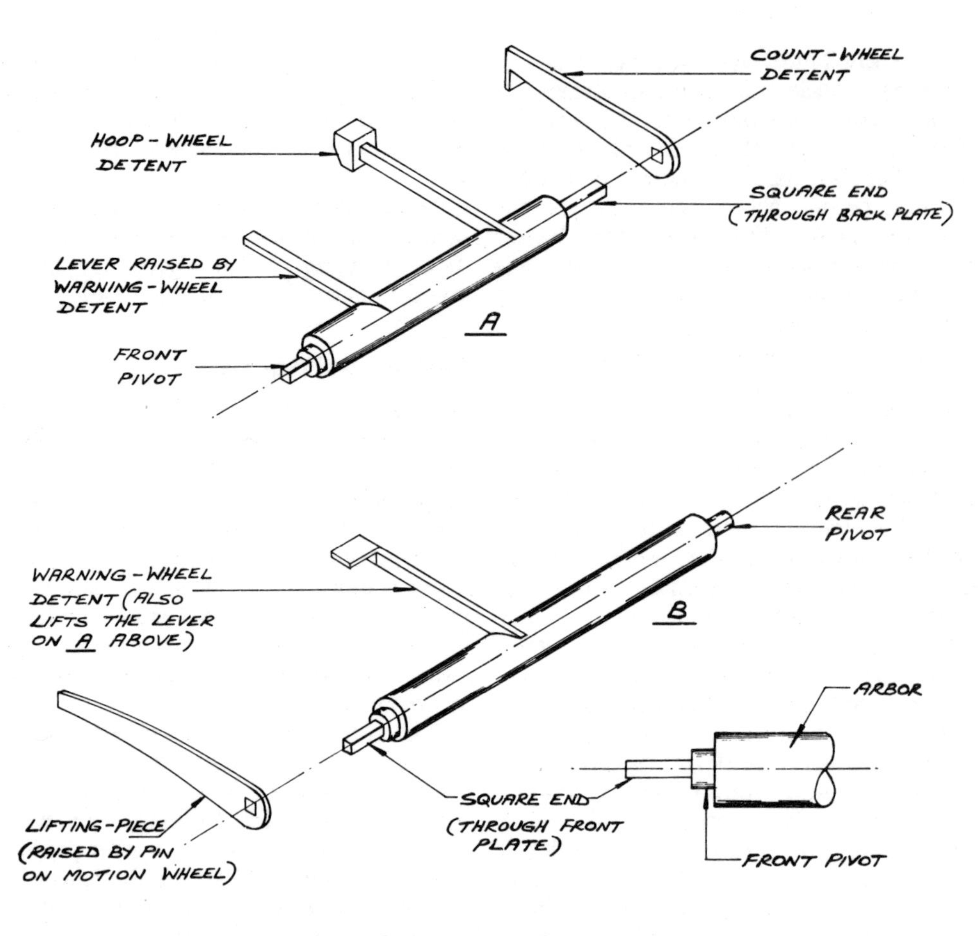

Figure 25

count wheel, and regulates the number of hammer blows.

The operation is as follows. When the train is still or locked, the count-wheel and hoop-wheel detents are down, ie in their respective grooves, and the lifting-piece is also down, all of them by gravity. As the lifting-piece is raised, its inside arm both raises the arm above it, and places a detent in the path of the warning-wheel pin.

There is some degree of play between the inside arm and the arm above it, because if the hoop-wheel detent was lifted too soon, the lifting-piece detent would not have risen, or lifted, far enough to hold up the warning wheel. Thus, the first thing to note regarding the setting is that the detent is actually placed in the path of the warning-wheel pin. Next the hoop-wheel detent, and with it the count-wheel detent, are raised out of their grooves, allowing both wheels to rotate. The warning-wheel pin is soon caught and held, leaving both levers poised.

When, at the hour, the lifting-piece falls in the normal way, it allows the hoop and count-wheel detents to fall also. Neither of these, however, can fall back into its groove, because each is prevented from doing so — the one by the fence on the side of the hoop wheel, and the other by the cam-like sections of the count wheel. The count-wheel detent now regulates matters entirely.

If you remove the lever from its squared end, and then operate the strike, the clock will strike only once because the hoop-wheel detent can now fall into its groove (space in the fence) after only one revolution of the wheel. The count-wheel detent, when remounted on its square, will prevent this all the while it is running on one of the cams, but directly it falls into one of the grooves it allows the hoop-wheel detent also to fall, and so lock the train.

It is most important, therefore, when assembling these levers and wheels to ensure that they are in the correct positions to permit the actions described.

The Hoop Wheel

77 The hoop-wheel detent takes the form of a small vertical piece with a flat end (see Figure 26), the lower part of which (marked B) is bevelled. When the lever falls, the part marked A is placed in the path of the fence or hoop on the side of the wheel. This hoop extends for about three-quarters of the circle only, and there is thus a gap into which the lever may fall to stop the wheel. When the lever is raised, the bevelled portion B ensures that the hoop is released by rising up over the fence to let the part marked C run freely on the hoop; and of course the detent cannot now fall until the count-wheel detent lets it do so. This occurs at the end of the strike.

Each cam on the count wheel is therefore of different length; for 1 o'clock there is no cam, because the lever is raised and falls again after one revolution of the hoop wheel. At 2 o'clock there is a very short cam, at 3 a larger one, and so on all round the count wheel up to 12 o'clock, the longest of all. On high-class clocks the cams are sometimes numbered 1 to 12, but not often on longcase clocks.

The count wheel or locking plate is rivetted concentrically to a toothed wheel which is mounted on a stud on the back-plate; it is located by a pin through the stud holding a friction washer, usually in the form of a fairly large triangle. The wheel is driven by a very small pinion squared on to the end of the great-wheel arbor of the strike train.

The arbor protrudes through the back-plate just enough for the little pinion to be located on it. The count-wheel plate, being of larger diameter than the wheel it is attached to, covers the small pinion and keeps it in place.

Adjusting the Strike

78 Various adjustments can be made to this comparatively simple system. Firstly, the play already mentioned may be varied by bending the lifting-piece lever up or down. Secondly, the count-wheel detent can be bent up or down; and lastly, the engagement between the little pinion and the toothed wheel of the count wheel can be varied to bring the cams to a position to suit the detent. The bent-down end of the count-wheel detent should ideally be in the same line as the radius of the wheel.

On some clocks the front end of each cam is bevelled, so that if the detent falls before reaching the effective edge of the cam, it is raised up, and the train is not locked in error. There are also instances where the count wheel is rivetted to a ring with internal teeth, rather than to a wheel; in these cases the small pinion engages the inside of the ring. More rare still is another type of count wheel which works in the reverse way to that described: here, the front of each cam is sharply bevelled to lift the lever to lock the train, and locking is done by pins, not a hoop.

79 In a 30-hour clock, the pins which operate the bell-hammer are usually on the great wheel, but are otherwise much the same as in an 8-day clock. Sometimes too there is an inside count wheel

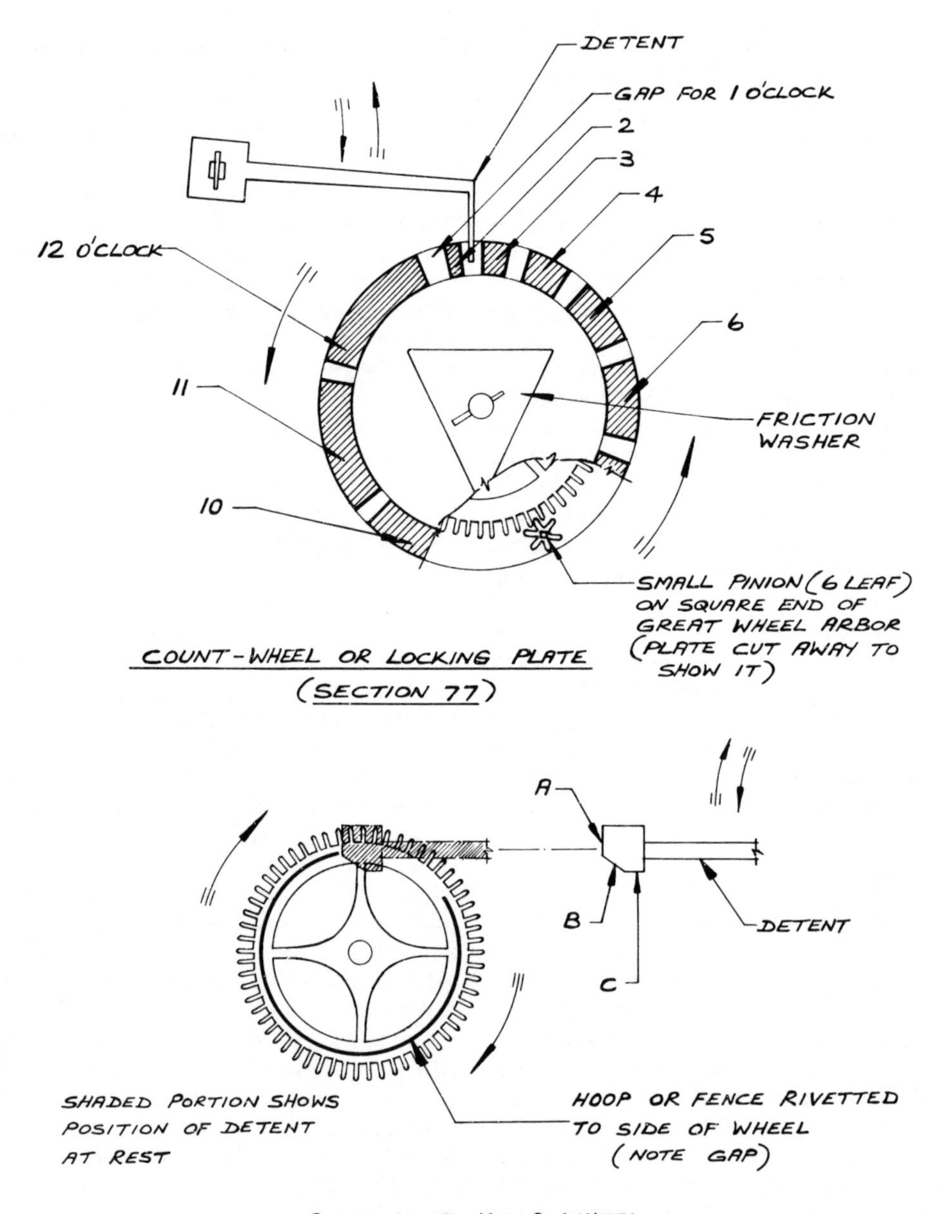

Figure 26

attached to the great wheel. There are also varieties of the system described. The principle, however, is the same, and any mechanically minded person will readily grasp the differences when confronted by a clock not quite the same as that described.

The 30-Hour Escapement

80 The escape wheel of a 30-hour clock is often larger and heavier than that of an 8-day; it revolves counter-clockwise in most cases, and consists of 45 teeth. The span of the anchor may also be different, but the operation of the pallets is the same. It is generally true to say that all the parts of a 30-hour clock are heavier and look clumsier than those of the 8-day. The drive wheels of a single-handed clock are thick and heavy, as is the twelve-point star wheel.

Incidentally, if the clock has a twenty-four point star wheel it will strike one at each half-hour; the slot in the count wheel is made wide enough for the detent to fall back into it after one blow of the hammer. This is very rarely found, but is worth mentioning here. A wide variety of adaptations are possible in executing the same principles of clockmaking.

Re-setting the Strike Sequence

81 It will readily be understood that a clock with a count wheel can be made to strike by lifting the count-wheel detent. This frees the hoop wheel but does not lift the warning-wheel detent so that the clock will strike no matter what hour is registered by the hour hand. The hand itself, or the strike sequence, will then have to be re-set. On many, if not most of such clocks one of the arms of the hoop-wheel detent is extended outwards to one side of the clock movement, and is grooved or drilled with a small hole to take a pull-cord; this, with a bead-weight, hangs down inside the clock case. A gentle pull on this cord causes the clock to strike, and it can thus be adjusted to strike the hour shown by the hour hand.

Alternatively, the hour hand is friction-mounted, and can be turned to point to the hour struck by the bell. Too much of this latter adjustment, however, can result in making the hour hand loose, which is not at all desirable.

There is another point worth mentioning. When winding a 30-hour clock, the pulling down of the chain to raise the weight causes severe friction on the arbor of the strike sprocket. It is wise therefore to raise the pulley chain with the left hand while pulling down the other with the right hand. This should *never* be done on an 8-day clock, but on a 30-hour clock it reduces the wear on all the parts concerned and makes for ease in winding.

82 Let us now look at the strike, or right-hand, moveable sprocket. If both sprockets are moveable the clock is probably quite old, and may originally have had two driving weights.

If the strike sprocket and great-wheel assembly is taken out of the clock, it will be seen that the great wheel is mounted integrally with the arbor but that the sprocket is free to revolve in one direction. On the side of the sprocket next to the wheel is rivetted a simple form of 'click' which catches in the spokes or crossings of the great wheel. This click can best be described as a circular inclined plane. Let us imagine a steel or iron wedge about 4 or 5in (120mm) long, one side straight and the other sloping like a right-angled triangle.

If we bend this round sideways in a circle, we get a circular ring with a step in it, ie one revolution of a screw. In practice the ring is solidly made, and that side of it opposite the step is rivetted to the side of the sprocket in such a way that the step may spring a small amount near to, or away from, the sprocket.

When this assembly is fitted close up against the great wheel the step slips under spring compression between two of the four crossings of the wheel. The sprocket is positioned laterally on its arbor, either by a steel pin through the arbor, or by a ring clip pressed into a groove specially cut in the arbor to receive it. The sprocket has therefore very little side play, and when it is rotated in winding the clock, the click described above is compressed inwards till the next spoke of the wheel rides over the step — and so on (see Figure 27).

After many years there will be considerable wear on the four crossings of the wheel, but if the whole click is either moved or re-rivetted, or replaced by a new one of a different-sized circle, then the step will contact a new part of the crossings.

On some very old clocks the writer has seen wheel crossings worn almost through, and where this occurs the play caused is excessive. By way of repairs, pieces of brass are occasionally found let in and rivetted to the spokes.

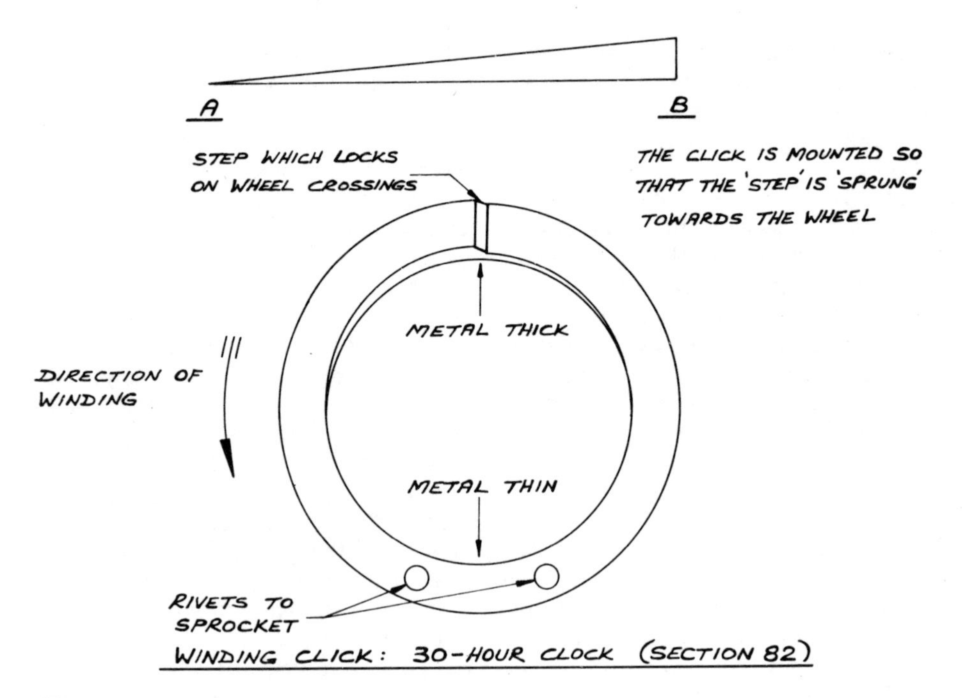

Figure 27

The step of the click should be as deep as the thickness of the wheel, because the weight of the drive is transmitted to the wheel only by this area of contact. It is obvious, therefore, that if the step engages only half the wheel crossing, the wear will be twice as hard as if the full width is engaged — the latter of course being the thickness of the great wheel. Later types of 30-hour clock have a spring-mounted hinged brass click which is simpler and more effective.

83 The counter-weight already mentioned has one important function and that is to ensure that the chain does not over-ride the pins of the sprocket. The chain consists of links, each at right angles to the next one, so that looking endways-on it appears to be cruciform. It is thus very easy to make certain that it is fitted straight and not twisted. Even so, the chain is light and will seldom hang true on its own without tension.

It is also possible, when the chain is slack, that one or two links may get caught broadways-on rather than lengthways; but the leaden ring goes far to prevent this, and also to provide sufficient tension to position each link entering the sprocket to receive the pin without slip.

On an old clock it is interesting to note the wear on the lead ring, caused by the chain as it passes through. After a while quite a deep groove is worn in the lead; and the ring in consequence will ride only in one position.

In a rope-driven clock, the sprockets are not shaped like those designed to receive links, but of course the spikes and the counter-weight are even more important to eliminate slip. Ropes are not popular nowadays because they fill the clock movement with fluff which has frequently to be removed. The normal way to join a rope is by splicing it, but clock ropes are especially woven and difficult to splice. The difficulty can be overcome by joining the two ends with a modern high-efficiency adhesive. If the rope ends are cut diagonally — leaving the surfaces elliptical — the glue has a larger surface to act upon, and if the join is carefully done it sets strongly enough to withstand the wearing effect of the sprocket pins.

Checking the back-cock pivot depthing with an extended runner of the depthing tool (Section 64)

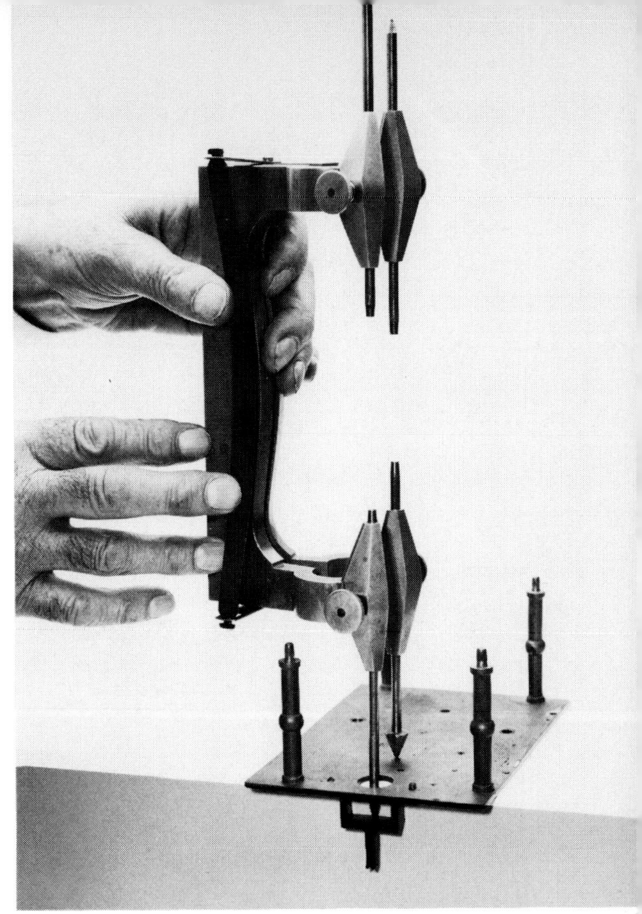

Photograph taken over a mirror, to show how a depthing tool may be uprighted in line with its own reflection in a clock plate (Section 37)

A brass 30-hour clock dial adapted to fit a spring movement. It has corner holes for mounting, and a winding hole beneath the centre. Note the name cartouche, the surrounding engraving, and the added date aperture. There are arrow-head hour marks within the quarter-hour circle, no minute marks, modest half-hour marks, and identical well-finished spandrels (Section 73)

Four pairs of longcase-clock hands (*left to right*): Scottish-style, decorated brass hands from a painted-dial clock; hand-made steel hands (not strictly matching) from a brass-dial clock; modern machine-stamped blued-steel serpentine hands; modern machine-stamped brass serpentine hands (Section 84)

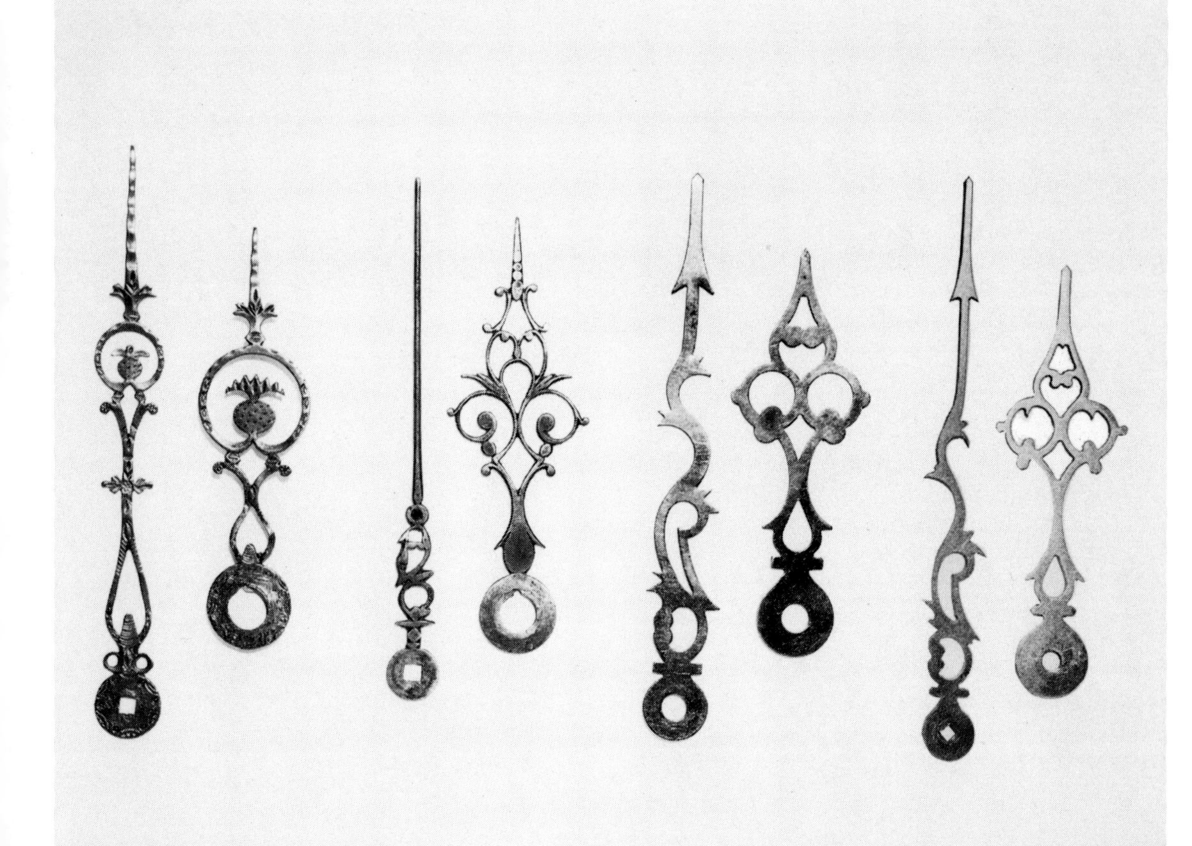

Repair of Hands

84 Your clock may be a century or several centuries old. During its long life its hands will have been set to the correct time a large number of times. The minute hand will have been pushed onwards by someone's forefinger, which is generally moist. Not only will the minute hand have rusted thin, but the dial itself will show wear, usually between the seventh and twelfth chapters. It will help, therefore, to describe how a clock hand, delicate at the best of times, may be repaired.

First of all clean up the hand, removing all rust, old lacquer, paint, etc, till the metal is bright. Do not use anti-rust solutions because most of these deposit a layer of zinc on the metal. With a lens, see if the shaft of the hand is cracked or made so thin by the cleaning that it needs repair. If so, either break it carefully, at the crack, or cut it cleanly in two with a fine saw or piercing saw — not with pliers.

Now make sure that the cut ends are dry and clean, and will fit neatly together. Fit the boss end of the hand in a vice on the bench. Fit the other first in a hand vice, which in turn is mounted in a second vice in such a way as to position the two pieces of the hand exactly as they would be when joined, but face down.

With a fine blow-torch gently heat the join after you have covered it with borax flux. When the borax has almost finished bubbling, put a tiny piece of silver-solder on the join, and have ready a length of steel wire with a point on it, with which to hold the little piece of solder in position. Now heat the join again until the solder 'runs'; as soon as it does, remove the heat and blow on the join.

Leave it to cool, and when cool, tidy up the sides and back — the front should not need it, except for a final polish to remove heat marks and any small traces of hardened flux.

Hands repaired in this way cannot readily or evenly be blued, so you will therefore have to enamel them with a quick-drying shiny black lacquer.

Bluing the Hands

85 The reason why the hands are blued, to that nice shade of steel-blue which looks so attractive, is that they are thereby made resistant to oxidisation, and also can be seen better against the background of the clock dial.

Before starting, make quite sure that the hands are clean, free of all rust, grease or moisture, etc, and the surface, especially the front surface, must be smoothly polished and burnished. This may be done with jeweller's rouge and oil, rubbed in a circular motion with a piece of boxwood. Alternatively, avoid all oil and use fine emery paper (No 500 or 600); after thoroughly cleaning all dust or fine emery from the hand, finish off with an oval burnisher.

At this stage the hand must not be touched with the fingers but handled with pliers or tweezers. A lasting blemish will be left if any fingermarks are put on the metal, and the whole process will have to be done again. Absolute cleanliness is essential.

Make an oblong tray about 8 × 4 × ¾in deep (200 × 100 × 20mm), and fitted with a handle about 8 or 9in (210mm) long. If you have no suitable tinplate, thin brass sheet or other suitable sheet metal, a fair substitute is to use the lid of a biscuit tin, but in that case there is no handle, which is a disadvantage.

Spread a layer of fine brass filings about ¼in (6mm) deep evenly in the tray, and mount it so that it can be readily heated from beneath. If no brass filings are available it is possible — but not really advisable — to use very fine clean sand. The operation should take place where there are no draughts.

With some tongs press the hand, face upwards, into the filings, none of which must be left on the top surface of the metal. The object is to spread the heat evenly over the entire hand.

Now apply the heat as evenly as possible up and down the tray from beneath and note the gradual changes in colour which take place on the shiny surface of the hand. Firstly it will turn to a light straw colour, then to a light blue, and finally to a nice deep blue. An even application of the heat is important, for the simple reason that the colour of the whole hand must be the same.

Directly this nice blue point has been reached – do not go beyond it – pick the hand out of the tray with a pair of tweezers, wave it in the air to cool it, and while it is still warm, wipe it over with a rag moistened with raw linseed oil. Remove surplus oil, and place the hand in a dust-free drawer to dry, preferably for at least twenty-four hours.

If you overheat the hand, or drop it on the floor, you will have to start all over again.

Making and Fitting a Replacement Rack Tail

86 On old longcase-clock movements it may well be found that the rack tail is damaged, or has earlier been repaired by soldering, to such an extent that it is now weak and unreliable. The rack consists of a small brass pipe mounted on a post in the front plate; on its inner end is the steel rack, and on its outer end the rack tail.

The rack has some thirteen teeth. It is positioned by the rack hook so that the gathering pallet may 'gather up' one tooth for each blow struck by the bell-hammer. The positioning of each tooth is governed by the setting of the rack tail. Thus it will be seen that the length of the rack tail is crucial; it must permit this 'one-tooth-at-a-time' collection.

The tail is generally made of hammered brass strip with a stout short pin at its outer end. Because it controls the amount by which the rack rotates on its post, the distance between the centre of the post and the centre of the pin in the rack tail is all-important.

Assuming you have removed the rack, and are about to remove the old broken tail, lay the rack inner side down on the bench, hold it firm, and gently file the rivetted metal where the rack tail joins the end of the brass pipe. Remove only the minimum of metal to permit the old tail to be pulled off. If there is any solder there, it will need to be melted off.

True up the shoulder on the brass pipe with a small file, but take great care here. Normally the piece would be put in a lathe and properly trued, but as the rack itself is firmly rivetted on the other end of the brass pipe this cannot readily be done. With care, however, you can make a perfectly sound job of it with a 'safe-edged' file.

With a piercing saw, cut out a new piece of thin brass slightly longer than the old tail and shaped to a curve, the arc of which roughly follows the outer contour of the hour wheel. This curve is only for ease of clearance of the wheel in assembling or dismantling.

Now, set a pair of dividers exactly to the radial distance between the first step on the snail and its twelfth step. This is the total distance that the rack will fall (in steps) between 1 o'clock and 12 o'clock. Make sure that the dividers take the measurement radially on the snail. Put the dividers aside for the moment without altering the setting.

87 Now lay the rack on a piece of clean paper, tail side up, and carefully mark the following three points:

1) The centre of the pivot hole
2) The position of the tip of the first tooth
3) The position of the tip of the twelfth tooth

These marks will be in the form of a triangle.

If there is a rack pin protruding rearwards, make a small hole in the bench to permit clearance.

Remove the rack and draw the two lines as shown in Figure 28, the point A being the centre of the rack pivot hole.

Take up the dividers and place them with one point on each of the two lines in such a way that DA=EA.

Any point on the circle GDEF with radius DA (or EA) will be the correct distance to work to for the rack-tail pin.

Now on the rack tail you have just cut out, mark the centre of the larger end, and from that mark, measure and mark the distance equal to AD on the smaller end. Drill small guide holes exactly on each mark. A fine centre punch should be used on the marks to make a little indentation

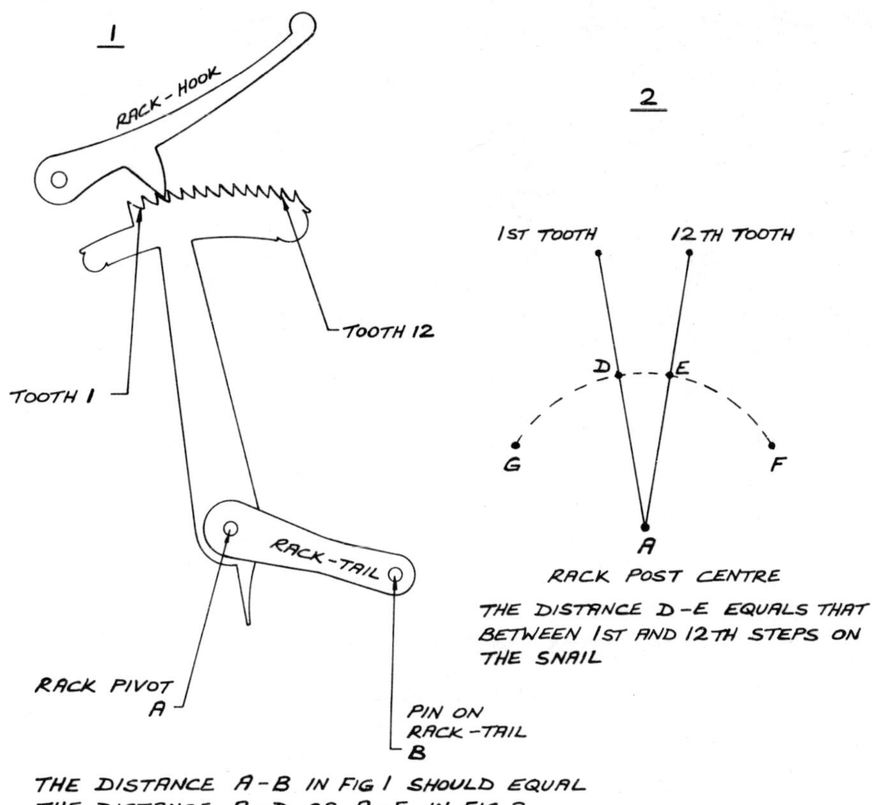

Figure 28

which will stop the drill-bit wandering.

We now have to make and fit a pin to the rack tail, and to fit the whole to the small brass pipe of the rack. Make a pin in the lathe like that shown in Figure 29, from a bit of iron or steel wire about one-tenth of an inch (3mm) in diameter; turn a shoulder on this pin so that its small end will fit tightly into the hole at the small end of the new rack tail.

Rivet the pin to the tail so that the bevel is to the right when you hold the tail horizontally in your left hand by its larger end.

The drawing also shows the pin rivetted in position, and the hole in the larger end broached out to fit the brass pipe of the rack. The joint of the rack tail with the small brass pipe must be sound and eventually rigid.

Drill out the hole at the larger end of the tail and broach it to fit the small brass pipe. At this stage the fit should be close and firm, but just loose enough to permit the angle to be altered under pressure. If the clock is dismantled, reassemble the strike train in the plates, and mount the rack, the rack hook, bridge and hour wheel on the front plate. Set the angle of the rack tail wider than necessary to clear the snail, and set the rack hook in the first tooth-gap on the rack.

Hold this firmly and push the new rack tail anti-clockwise till the pin rests on the first step of the snail — the latter having been positioned accordingly. Try the run,

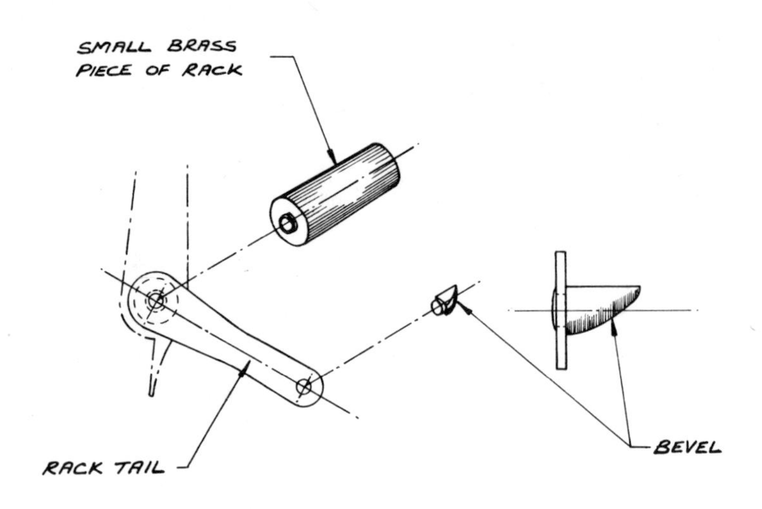

Figure 29

taking care not to disturb the rack-tail setting. The one tooth should be gathered up and the train locked again. Now try other steps of the snail, and if all are correct, rivet the tail rigidly on to the shoulder of the small brass pipe so that it cannot move at all in relation to the brass pipe.

When trying out all the steps of the snail in this way, mount the motion wheel and the lifting-piece in addition to the other parts just mentioned, and operate the parts by manually turning the motion wheel anti-clockwise. You should now have a good sound rack tail. It is still essential, however, to set the tail-pin bevel to meet the bevel on the snail in such a way as to ensure that the rack tail is pushed aside by the snail in the event of the hands being turned forward when the rack is 'down', or when the strike is 'expended' while the going train still runs.

When rivetting the larger end of the rack tail on to the brass pipe, you will have somewhat distorted the centre hole of the pipe which will not now fit freely on its post. With a finely tapered broach ease the hole out from the inside, to its correct size and no more. There should be complete freedom for the rack to rotate, but no side play at all. Next, re-set the rack spring so that it guarantees the regular fall of the rack. It is unwise to overdo the spring pressure as this only causes the pin on the rack tail to hit the snail unduly hard whenever the rack falls; damage or undue wear may ultimately result. Also, there is almost certain danger of the pin wearing, so causing the bell-hammer to strike the wrong number of blows, or at worst causing the strike to fail. Sometimes if the rack is slightly out of position, the tooth of the gathering pallet may come to rest on the tip of one of the teeth of the rack, and this will stop the train altogether. Trial and error on each step of the snail will ensure good action.

88 The rack-tail pin moves, as we have seen, in a circular path, and this means that the pin contacts the snail at a

slightly different spot on its perimeter at each step of the snail. Unless, therefore, this perimeter is circular you may get a slight variation in the setting. For example, the strike may work perfectly up to the eighth hour but strike 8 again at the ninth step. A mere touch with a file on the pin exactly at the point of contact will remedy this.

If, however, having struck correctly up to 8, the next strike is 10, then the tail will need adjustment. Again, by trial and error, with the parts moving slowly, see how much space has to be taken up. This may be done carefully by twisting the tail to bring the pin into closer contact with the ninth step on the snail.

Try all the hours till the strike is perfect, for much time spent now on this will save hours at some later date.

Fitting a New Gathering Pallet

89 The rack teeth may well have worn irregularly in the course of a century or so; and this means that the gathering pallet may move the rack a different distance each revolution. The pallet itself may also have worn, though the difference is usually small. Again the pallet may move the rack by more than the prescribed one tooth. If this movement is too great – eg if 3 is struck at the sixth step – it means that the pallet tooth is too long. Study of the pallet tooth shows that its tooth (or leaf as in a pinion) is only half a leaf, being flat on one side and curved on the other. Its operative tip is radial, and the flat side should never be filed; if therefore the pallet tooth is too long, shorten it by filing only on the curved side (see Figure 30). If the movement of the rack is more than one but less than two teeth, the rack hook usually falls back to the correct setting. The rack hook will slide up part of a tooth and then fall back again, so that only one tooth is actually gathered up. If this happens only in one or two cases out of the twelve, it may mean that the rack teeth are not regular. If it happens at all twelve steps, either the pallet tooth is too long or you have got your measurement wrong; ie the exact radial distance on the snail between the first and twelfth steps. If this is badly out, you will have to start all over again, but before doing so, make quite sure that the steps of the snail are correctly set – they probably will be – and that the rack teeth are even and regular.

The pallet on your clock will probably be all right, but if it is too short, and will not move the rack far enough to gather up the teeth, a new one will have to be fitted. (Much has already been said about the pallet tail earlier on, see Sections 39, 42 and 43.) If you have to fit a new pallet you will need to buy one from a materials dealer. He will ask you for what sort of clock it is required, and whether you want a left hand or right hand. You want it for a longcase movement, but as to right or left you will not know. Think back to the rack and, in imagination, as you face it on the front plate of the clock, there will be a pin on its left side which catches the tail of the gathering pallet, so locking the train at the end of each strike. This pin may either protrude from the front of the rack or from the rear; the pallet tail will

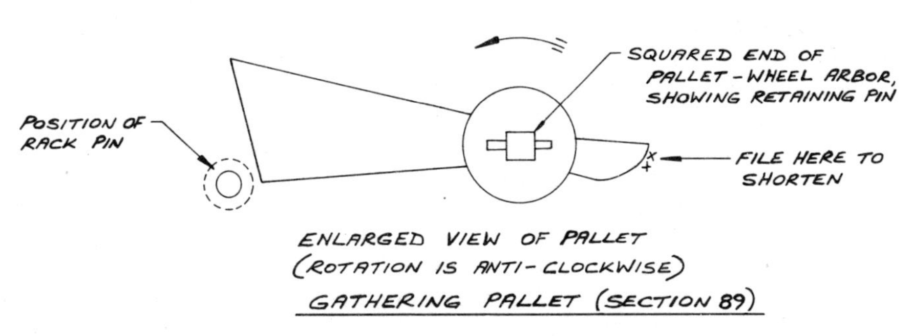

Figure 30

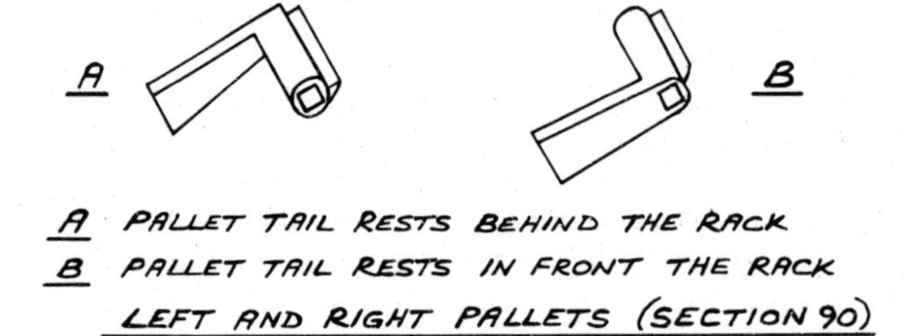

Figure 31

accordingly work on a different side of the rack, and also, of course, of the pallet body. This is best seen in Figure 31. In the past, one could buy gathering pallets for a few shillings, but now they may cost two or three pounds or so. When you obtain one it will be a small casting with a minute hole drilled through the body of it. This hole finally has to be made square and tapered in order to fit the arbor on which it is mounted. Both the leaf and the tail will be too long and will need filing to shape as in Figure 30.

An important point to remember is to ensure that the pallet tail just clears the rack pin when the rack hook is in the first tooth of the rack, ie is in the 1 o'clock position. For this job you will need some square-section needle files from the materials dealer. They vary in length and thickness, and it is virtually certain that you will break one or two of them before you have finished. (The skilled reader may smile ruefully at this, remembering his early days.)

Enlarge the hole in the new gathering pallet to about half the diameter of the pallet-wheel arbor, and with the square needle file, start to file four corner nicks in the hole from the inside, ie the side of the pallet next to the front plate. Copy the old pallet, gradually and evenly enlarging the square hole, making it fit the end of the pallet-wheel arbor as you proceed. When the hole is squared, some makers drive the file quite forcibly once or twice into the hole and out again in reverse; but this, one feels, has a tendency to split the pallet. Patience and skill can achieve a good fit. The pallet needs to be pushed firmly on to the squared arbor far enough to allow its tiny retaining pin to be pushed into its hole. It must be a good firm fit. The taper fitting permits this tightness, a fact readily understood; so that it is necessary to make sure that the pallet retaining pin is fitted close against the pallet so that the latter cannot retract on the taper square, and so loosen. Any play could be taken up by inserting a small shim washer beneath the retaining pin, but this is difficult because it is so small, and should not be necessary. Sometimes the pallet is held in place by a very small threaded nut, the end of the squared arbor being threaded to receive it. This ensures a good fit provided that the little nut is tight.

90 Now that the gathering pallet has been fitted to its squared arbor, attention must be paid to its clearance of the rack teeth, and also to the sideways movement of the rack, as stated earlier. Assuming that the rack is not damaged, and that all its teeth are equidistant from the rack post, test the fall of the rack in all positions of the gathering pallet.

See that the leaf of the latter does not foul the rack hook or lifting-piece in any way, and that the gathering-up is as described earlier (Sections 42-3). If all is well, you may be confident that you have done a good job. At any rate, if you have not you will know how to correct anything you have so far done.

If, however, any of the rack teeth touch the body of the gathering pallet, the body must be carefully filed to give adequate clearance — always assuming, of course, that the rack is true and not bent.

91 Before leaving the subject of the rack it is worth mentioning that the plane of the rack must be the same as that of the rack hook. This can be tested by lifting the rack hook just clear of the teeth and moving the rack back and forth. Each tooth should pass exactly under the hook in the same line, and there should be no side play either in the rack or the hook.

If there is any undue play in any lever which pivots on a post (eg rack, rack hook, lifting-piece), it may be lessened by hammering the inside gently while the small brass piece is set firmly in a stake hole; and if the collet is closed too much by this, open it with a round broach, using only as much force as is necessary for free rotation.

92 In the case of the rack hook for instance, if there is any play in the collet, wear is 'encouraged' by the hook being moved first right, then left, at each revolution of the pallet. It would take a long time to affect the setting or position of the rack teeth in relation to the pallet tooth but eventually this would happen. If on the other hand the rack hook is laterally firm, it so positions the rack each time that its teeth are evenly and regularly gathered up. Similarly, any play in the collet of the rack itself will affect the setting, as can readily be seen.

It is not a waste of time to seek perfection in these matters; indeed there is very considerable satisfaction to be gained by doing such jobs properly, especially as we are not always working with the best tools.

EPILOGUE

We have come a long way since beginning to restore our grandfather clock, and have learned enough to become interested in further horological exercises. This booklet has but scratched the surface of the subject, dealing with a large type of clock that is comparatively easy to restore or repair. There are myriads of mechanical watches, electric clocks and watches, bracket clocks, wall clocks, regulators, quartz clocks and watches, and so on — to say nothing of church clocks, stable clocks and other forms of turret clock.

Some very comprehensive (and expensive) books on horological subjects are available, many of which may be obtained from the local library. To aid you in choosing them, a short bibliography follows the glossary. It includes all sorts of books, some technical, some historical, some aesthetic — even one novel. Study your clocks, practice working on them, and may all good wishes go with you.

John Vernon
Ringwood 1982

GLOSSARY

Anchor The pallets of a recoil escapement, sometimes also of a dead-beat escapement; name derives from the shape of the pallets

Arbor An axle or rod on which are mounted wheels, pinions, etc, usually with a pivot at each end

Arch An arch the full width of the clock case (*See also* Break-arch)

Arabic Numerals Ordinary figures, in the form we write today. Used to denote the number of minutes and seconds, and date numbers. Roman numerals, as used for the chapters, are never used for the minutes, etc. The size of the Arabic numbers of the minutes can be a clue to the age or period of the dial of a longcase clock

Astronomical Clock A clock showing the comparative motion of planets, etc; often with a twenty-four hour dial. Seldom found in the domestic range of clocks

Back-board The wooden back of a longcase clock, usually made of elm wood. It extends upwards to form the back of the hood

Back-cock The 'bridge' made of brass and mounted on the back-plate which carries (1) the rear pivot of the pallet arbor, and (2) a slotted extension which supports the pendulum

Back-plate The rear plate of a clock, to which the pillars are rigidly rivetted in nearly all longcase clocks

Banding Using bands of inlaid veneers on a clock case, such as for the strip round the edge of a door, where the grains are generally at right angles to one another

Barley Twist The style of the twisted columns at the side of the hoods of longcase clocks

Base Plinth

Base-plate Another name for the dial plate of a brass-dial clock with separate rings and spandrels

Beat A pendulum is 'in beat' when it swings evenly, ie when the ticks are evenly spaced

Bob The weight on the end of a pendulum rod. For false bob, see False pendulum. The usual shape of the bob of a pendulum is known as lenticular

Bolt and Shutter Part of the mechanism for keeping the power going while the clock is being wound. On a longcase, the shutters cover the winding holes, and must be removed before the clock can be wound. Such action brings a subsidiary spring into play, to maintain power (*See also* Maintaining power)

Bolts *See* Seat bolts

Brass Block This block rides in the fork or hoop of the crutch, the lower end of the suspension spring of a pendulum being hinged to it. The top end of the pendulum rod is screwed into the lower end of the block, which receives impulse from the crutch

Break-arch An arch of half-circle size or less on a clock dial or its case, which is narrower than the dial or case

Bridge A support, generally of brass, with two angled ends each of which is screwed to a plate, usually carrying part of the motion work (*See also* Back-cock)

Broach A long, thin, tapered tool with five or six 'sides', used for opening up holes in metal plates. When using cutting broaches, careful rotation cuts the edges of the hole. When using round broaches the edges are polished hard

Bun Feet Rounded bun-shaped wooden feet, on which some longcases stand

Bush or Bouchon A small brass bearing in which a pivot revolves. New bushes can be inserted in the plates where the original pivot holes are worn.

Butting A leaf of a pinion or a wheel-tooth may 'butt' or come up against another. This is bad, showing either wrong depthing or a wrong size of pinion. Butting will stop the clock
Calendar Aperture A hole cut in the dial plate of a clock to show the date or day of the month
Calendar Ring Date ring
Cannon Wheel The wheel on the front plate, the hollow arbor of which carries the minute hand, and is driven by a friction washer from the centre arbor; another name for the minute wheel
Capitals The tops of the columns on a longcase hood, usually in brass
Centre Seconds Term used when the seconds hand of a clock runs concentrically with the other hands. Sometimes found on longcase clocks of high quality with dead-beat escapements. Centre seconds hands are carefully balanced about their centres (*See also* Sweep seconds)
Centre Wheel The wheel, usually second in the going train, whose arbor rotates the minute hand and is the longest arbor in the clock
Chapter Ring The circle showing the hours and minutes. The term usually refers to a separate flat ring with four 'feet' for mounting on the dial plate
Chapters The hour numbers — normally in Roman numerals but not always
Cheeks The wooden uprights inside the case of a longcase, on which the seat-board rests
Chime Clock A clock which chimes the quarter-hours as well as normal striking
Chops The two sides or pieces of metal on the back-cock which grip the end of the suspension spring of a pendulum
Cleaning Fluid Special fluid for dilution, used to make a solution for cleaning clock or watch parts; obtainable from materials suppliers
Click A pawl running over ratchet-wheel teeth, or barrel teeth, in a longcase clock, in order to prevent the return of the ratchet. The pawl is sometimes fixed rigidly in a spring-driven clock, but is always free in the winding assembly, being held in contact with the ratchet by a light leaf-spring
Click-spring The light spring which tensions the click (*see* Click)
Clock Said to have been derived from the German *glocke*, a bell, from ancient times when clocks had no dial, merely bells that monks would ring at certain times
Cock A bracket with one foot. It usually carries the pivot of an arbor, the other end of which is carried in the plate on which the cock is mounted
Collar A metal ring — open like a collar — used to encircle something, holding it by friction. A push-on ring to hold a clock hand to which it is soldered. A spring-tension ring lying by its own compression in a groove cut for it in an arbor to position a wheel otherwise free on that arbor
Collet A small brass hub to which a wheel is fixed when not rivetted to a pinion. The pallets are generally mounted on a collet. If the wheel or pallets are rivetted or brazed to a collet, and the latter soldered to an arbor, the collet may later be easily moved and adjusted as required. This is a point worth remembering
Columns Another name for pillars, ie turned rods which hold the plates together. In a longcase clock the columns are usually rivetted into the back-plate permanently. They are pinned or latched to the front plate so that the latter can be removed.

The word columns is also used to describe the side columns on the hood of a longcase clock (*See* Corinthian columns)
Comtoise Clocks French wall or long-case clocks common on the Continent. They are also called Morbier or Morez clocks from those villages in the Comtoise district where they are made, Franche Comté
Contrate Wheel Very rarely found on English longcase clocks after about 1690. It is a wheel with teeth on the side of the rim, ie at right angles to the plane of the wheel. Contrate wheels drive crown-wheel pinions in bob-pendulum movements, and sometimes moon-work
Convex Moulding *See* Moulding
Corinthian Columns Fluted columns used decoratively on the hoods of long-

case clocks. They normally have brass capitals at each end (*See also* Barley twist)
Corner Pieces Spandrels
Counter-sink A small tool or bit for a drill, with which the entrance of a hole may be bevelled – to receive a screw-head, for instance, or to permit clearance for an arbor. The counter-sink has a conical steel head which is radially cut to form teeth. When placed upright in the hole in the wood or metal, and rotated under slight pressure the cone will cut the edges of the hole to the depth required
Counter-weight The small weight, usually a ring made of lead, which maintains the tension of the rope or chain used to drive a 30-hour longcase-clock movement (the name has other meanings not relevant here). Also, the tail of a clock hand may be so weighted to maintain balance
Count Wheel Another name for the locking plate which governs the number of strokes when a clock strikes. Various forms are found on longcase-clock movements, some inside, some out, and all can get out of sequence with the hour hand
Cresting A term used for the upper decoration of the hood of a longcase clock
Crossings The spokes of a toothed clock-wheel. The process of making such spokes from the solid 'blank' is known as crossing out
Crown Wheel An escape wheel like a crown, with teeth sloping on one side and vertical on the other, for operating the pallets of a verge escapement. The teeth are at right angles to the plane of the wheel
Crutch A slotted wire arm fitted to the arbor which bears the pallets. The block on the pendulum rod is located in the slot of the crutch, and so receives impulse imparted to the pendulum. The latter can be put in beat by bending the crutch till the tick is even (*See* Beat)
Date Hand There are two types, both found on longcase-clock dials. One is central, ie concentric with the clock hands, indicating the dates on an appropriate circle of figures inside the chapter circle; the other is an ordinary small pointer on a dial above the VI or, more rarely, in the break-arch
Date Ring A large circular ring rotating behind the dial plate of a clock so that through an opening in the dial the day of the month may be read. The ring is engraved with numbers 1–31, and is almost always silvered. The edges of the opening in the dial are bevelled, so that the ring may be moved on manually when a short month occurs. Otherwise the ring is moved once per day by the clock
Date Wheel Part of the motion work which operates the date ring by means of a protrusion which catches one tooth per day
Dead Beat A type of escapement with no recoil whatever. The best dead-beat escapements have jewelled pallets. Steel pallets tend to lack oil after a while, or it oxidises or thickens, which can seriously affect the operation, and hence the time-keeping
Depthing Tool A special tool whereby the correct depth of engagement between wheel-teeth and pinion leaves may be accurately set and, by means of compass-like points, transferred to a clock plate, so that the gearing mesh is accurate
Detent That part of a lever or hook whose function is to hold up some operation of the clock train until the time is ripe to release it. In a strike train, for instance, the detent or detaining lever on the lifting-piece holds the warning wheel till the minute hand points directly to the 60-minute mark. A chime train has similar types of detent
Dial The entire face of a clock. Any circular subsidiary, either on its own or within the main dial, eg seconds dial, date dial, etc
Dial Plate The brass plate fixed to the front plate of a clock, on which are mounted the various rings, spandrels, lunar work, etc. The whole face of a painted-dial clock (*See also* False dial)
Drop The free fall of an escape-wheel tooth on to the pad of a pallet; the sound of this is the tick-tock of a clock
8-Day Clock A clock which runs for eight days, its barrels rotating twice per day. Eight-day longcase clocks have seconds pendulums, escape wheels of thirty teeth, and generally a seconds dial
Equation Clock A clock which shows

mechanically the difference between sun time and mean time — called the equation of time

Escapement A mechanism to control the speed, and count the rate of run-down of a clock. An escape wheel is held up tooth by tooth and released similarly by various devices. On a longcase clock the control is by pallets. There are a great many forms of escapement, the best of which interfere with the pendulum as little as possible

Escape Wheel On the ordinary 8-day longcase clock this wheel has 30 teeth, operating an anchor or recoil escapement. A 30-hour clock usually has a heavier wheel with more teeth

Escutcheon A small, often decorative, piece of brass or metal plate surrounding a keyhole; it literally shields the surround of the hole

False Dial (or false plate) An inner rectangular cast-iron frame-like piece to enable a painted dial to be fitted to the front plate of a clock. If a false dial is found on a brass-dial clock it should be at once suspect. Examine it to see if another movement has been fitted to the brass dial or vice versa

False Pendulum Seldom found in longcase clocks made after about 1685. A small brass disc connected to the verge of a verge escapement clock; it may be seen through a slot in the dial

Fence The curved cam fastened to the side of the hoop wheel in a 30-hour longcase clock. Its function is to lock the train (by means of trapping a detent) when the lifting-piece is down, after the clock has struck the hour. There is sometimes another kind of fence on one of the motion wheels which operates a lever in a clock with pump action strike levers (*See also* Pump action)

Fire Gilt Another name for mercury gilt

Fly A small air brake — the last piece of the strike train

Fret Either wood or metal fretwork in the sides or front of the hood, sometimes silk-lined to prevent dust entering. Frets permit the sound of the bell to be heard more clearly

Friction Fit Used here to describe the fit of a slotted pipe over the end of an arbor, or of a ring or collar on a wheel pipe. In some instances termed a push-on fit

Friction Washer A washer fitted against a wheel which is not fixed to its arbor. Side pressure on this washer presses it against the wheel, the resultant friction causing the wheel to rotate whenever the washer rotates. In clocks it is used in several ways: 1, on the centre arbor behind the minute wheel, to drive the minute wheel and to enable the hands to be set; 2, on the hour wheel, so that the hour hand may be moved in relation to the minute hand, to obtain an exactly correct setting on the dial; and 3, on both great wheels to permit winding

Front Plate The foremost of the two plates of a clock movement. On some clocks it is mounted in two halves for ease in dismantling and assembling one train without disturbing the other

Gathering Pallet A one-toothed pinion on the squared end of an arbor, positioned so as to gather up the teeth of the rack in rack-striking clocks. Clocks with internal racks have a slightly different arrangement but still use the one-toothed pinion

Gilt *See* Mercury gilt

Going Train The train of wheels which, running continuously, operates the time indication of a clock

Graining A method of treating a metal surface, such as a dial plate or a chapter ring, either for a certain finish, or for later silvering. After thorough cleaning of the plate or ring it is carefully 'grained', all one way, with a fine pumice block or emery block. This gives a pleasing appearance, especially when silvered

Grandfather Clock A late nineteenth-century name for the longcase clock, taken from the words of a music-hall song of the time, referring to 'my grandfather's clock'. The name stuck and is so well known that this note is almost superfluous

Grandmother Clock A development of the above: a much smaller clock, though similar in other respects. A smaller one still has, in modern times, been termed a 'granddaughter clock'. Real antique grandmother clocks are rare indeed. They are usually not more than 6ft high

Great Wheel The largest wheel in a

clock's train of wheels. Attached to the barrel in an 8-day longcase and to the fusee in a bracket clock

Half-hour Marks The half-hour divisions on a brass or silvered chapter ring (*See also* Quarter-hour ring)

Hands Sometimes, albeit archaically, referred to as fingers. They are of infinite variety and run roughly to fashion, as did spandrels. Hands, if original, are one of the many details to consider when dating a clock. They all indicate something on a dial

Hand Collet The small saucer-shaped washer fitted over the minute-hand boss and pinned to hold the hand in place

Hammer The little hammer which strikes the bell on a clock

Hammer Tail A projection from the arbor which carries the hammer. It is motivated by the pins on the pin wheel, ie lifted and released at each stroke

Holding-down Bolts (or seat bolts) The two bolts which fasten a clock movement to its seat-board

Hood The glass-fronted top-piece of the case of a longcase clock

Hoop Wheel The wheel in the strike train of a clock, generally a 30-hour clock, which has a circular fence attached to one side for about three-quarters of its circumference. A detent is set so that it may run or travel on the surface of the fence when the wheel rotates, till it falls into the gap and so arrests further motion, ie the train is locked

Hour Circle The circle showing the hours — just beneath the numerals — on a chapter ring or on a painted dial

Hour Hand The shorter and wider of the two hands of a clock, often finely decorative. Its length is correct when its tip just reaches the hour circle

Hour Wheel A wheel of seventy-two teeth in the motion work, on the pipe of which is mounted the hour hand

Impulse The term used to describe the escapement's action in giving power to the swing of the pendulum (or to a balance wheel in a watch)

Inlay A form of veneering where pieces of wood, bone, shell or metal are let into a surface direct or among other veneers. There are many forms including Buhl work, intarsia and marquetry

Intarsia The earliest form of inlay

Intermediate Wheel Another name for motion wheel

Jewelled Pallets In a longcase, the sapphire or gemstone insets let into the pallet arms to form pads on which the escape teeth fall. Found only in high-quality longcase clocks and nearly always in regulators

Journal A term not generally used in horology. It refers to that part of an arbor which runs or revolves in a bearing

Jumper A spring device for holding a star wheel or date disc, moon wheel, etc, exactly in place after movement, in order that the wheel's relevant tooth may be in the right place for engagement at the next movement. A star wheel, for instance, is pushed about three-quarters of its travel by a pin, by which time the jumper takes over, pushing the wheel the rest of the way and holding it there. Moon-wheel jumpers and others may also be 'non-return' devices

Lacquer Coloured or clear varnish used to decorate clock cases. High-quality longcases have very fine oriental lacquer work, the cases having been sent by sea to the Far East for treatment and then re-imported.

Also, clear spirit varnish, or the like, used to coat metalwork to keep it bright, free from oxidisation and tarnish

Lantern Clock Said to be the forerunner of the English longcase clock. The clock is made mainly of brass, with four corner pillars, and strips of brass wedged between the top and bottom plates to hold the trains. The bell, held in a four-armed strap of brass, is mounted on the top of the clock, with a finial above it, to match the four corner finials. The name is said to have come from the French word 'latten', meaning brass, but the clock does resemble an old-type lantern — or so some people think

Lantern Door A glass window, usually round, set into the door of a longcase clock to enable the pendulum bob to be seen

Latched Plates The plates of a clock movement are sometimes latched, instead of pinned, together. One end of a hook-like

flat latch, hinge-rivetted at one end to the front plate, slides by rotation into a groove in the top of a pillar to hold the plate in place. The latches are turned back to release the plates. Not often found after the middle of the eighteenth century
Lifting-piece The lever which is lifted by the motion work in order to prepare the clock for striking or chiming, and to release the strike at the correct moment
Lifting-pin On a longcase clock, the pin on the motion wheel which raises the lifting-piece to make the clock strike
Line of Centres The line joining the centre points of two axes, ie between wheel and pinion centres (*See* Section 37)
Lines The cables, cords or gut lines carrying the pulleys and weights of a clock. One end of each line is attached to the barrel and the other via a pulley to the seat-board. A weight of 12lb would effect a gravity pull on the barrel of 6lb (friction omitted). The use of a pulley also halves the distance that the weight would travel if suspended from the barrel without one
Locked In horology this means that a moving train of wheels has been brought to a stop. The train is 'locked'
Locking Plate A way of regulating the number of strokes made by a hammer on a bell. Also called count wheel. There are various forms but they are all on the same principle.
Longcase The 'correct' name for a grandfather clock – correct in that the name is some hundreds of years older
Lunar Dial A dial or revolving plate, usually in the break-arch of a longcase clock, denoting the phases of the moon. Its operation is known as lunar work. In olden times it was important to know if there was a moon if you had to go out at night when footpads were active. The lunar dial on a longcase clock rotates once in two months (*See also* Moon wheel)
Maintaining Power A method whereby the clock is kept going while it is being wound up. The winding of a longcase clock, for instance, means that the weight is held by the key during winding, and not by the great wheel. Maintaining power applies power for a short time only, until winding is complete or a minute or two thereafter. Found only on high-quality and precision clocks (*See also* Bolt and shutter)
Marquetry Inlay of intricate patterns and often of different colours or shades of wood veneer. Some marquetry on clock cases is supremely excellent
Matting The matt treatment of the centre of a dial plate within the circle of the chapter ring. The art is now lost. It produced a fine granulated effect like modern sand-blast, and it looks especially well within silvered chapter rings and silvered seconds rings
Mercury Gilt A fine form of decoration of brass or similar metal with gold. The gold was dissolved in mercury to form an amalgam, which was used to transfer the gold to the brass by heat; this evaporated the mercury, leaving the gold. The process was extremely toxic and many an operator got mercury poisoning
Minute Circle The circle on a chapter ring immediately outside the chapters
Minute Hand The long hand of the two main hands of a clock. It indicates the minutes and should be of a length just to reach the minute circle (*See* Hands)
Month Clock A clock which runs, usually for 4×8 days=32 days. An extra wheel in the going train achieves this, and means that such front-winding clocks have to be wound anti-clockwise.
Moon Wheel A large painted disc indicating the age of the moon, and mounted in the break-arch of the dial plate. More rarely it may be found in the upper segment of the dial within the chapter ring, in which case there is no seconds dial
Moon-work The method of rotating the moon wheel by levers operated by a cam on the hour wheel of the clock (*See* Lunar dial)
Motion Work The gearing reducing the minute-wheel rotation to that of the hour wheel. Motion work is between the front plate and the dial plate of the clock. The date wheel is also driven by the motion work
Movement The mechanism of a clock (horologists snort at the word 'works')
Musical Clock A three-train clock which plays tunes at intervals, some at each hour, others at longer intervals – in a

longcase clock nearly always on bells. If the same note is repeated rapidly in the tune two separate hammers are needed (*See also* Organ clock)

Moulding The shaped wooden or curved pieces joining the plinth to the trunk, and also the hood support to the trunk. Until about 1700 many longcases had convex moulding, but around that time mouldings became concave. If you see an old case or complete clock with convex mouldings in a saleroom, buy it — even if you have to borrow money to do so

Name Plate A silvered brass plate, cartouche or roundel attached to a dial plate, with the name of the maker of the clock, or sometimes the name of the retailer of the clock, on it. It is all too easy to make and attach a name plate engraved with any name one likes. Where it shows a maker's name, especially one of some fame, it may be suspect; if the method of attachment is obviously different from that used for the other fittings, it could have been put there to deceive. A retailer's name, however, could well have been added, perhaps decoratively or in the course of advertising

Non-ferrous Parts In our context, all parts of a clock which are made of a metal other than iron or steel

Oil Sink A small keen-edged depression on the outside of a pivot hole. It retains a little oil in its hollow and prevents the oil from leaving the bearing because oil will not 'creep' over a sharp edge. It is unwise to make deep oil sinks in plates which are thin because the bearing surface is reduced and uneven wear on the pivots will result. Only oil of the highest quality should be used and sparingly at that. Oil which collects dust in a pivot hole is the clockmaker's worst enemy

Organ Clock A musical clock which plays tunes on a small organ. Seldom if ever found in a domestic longcase

Pad Another name for the surface of a pallet in a recoil or verge escapement

Painted Dial A term used for factory-made dials supplied to clockmakers, fully colour-painted or stoved, leaving the chapters to be painted in by the clockmaker. Such dials supplanted brass dials, either because they were easier to read or because they saved the average clockmaker a great deal of work. In time, they were also cheaper. When supplied from the foundry they were fitted with false dials. Often referred to as white dial

Pallets Hard steel 'flukes', as it were, in an anchor escapement, and little plates on the staff of a verge escapement. The teeth of the escape wheel fall in turn on to the pallets, pushing them away and so imparting impulse to the pendulum via the crutch

Pallet Staff The arbor on which the pallets are mounted, and which, in a longcase, carries the crutch at its rear end

Pallet Wheel The third wheel in an 8-day striking train. The front end of its arbor protrudes through the front plate, and is squared to carry the gathering pallet; hence the name pallet wheel

Passing Strike *See* Trip strike

Pawl In a clock, a pawl is called a click. It is shaped to prevent the return of a ratchet or rack after it has moved in any direction. When a clock is wound, the click passes over the teeth cut in the barrel end-plate, and by preventing it unwinding, ie its return, transmits the pull of the clock-weight to the great wheel

Pediment The triangular top of an architectural case. Where the line is broken on top, it is a 'broken pediment'. Pediment, more loosely used, means that part above the cornice in a clock case. Also portico

Pegging Out The process of cleaning pivot holes with pegwood

Pegwood Small sticks of dogwood used for cleaning out the pivot holes of a clock. The pointed end of the stick is rotated in the hole under light pressure, removed, scraped clean and re-inserted till the wood remains clean

Pendulum The part of the clock which regulates the time. A swinging weight on a rod or shaft hanging by a suspension spring from the back-cock of the movement. The pendulum oscillates, and the shorter it is the quicker it swings, and thus regulation, faster or slower, can be achieved

Pendulum Aperture A curved slot through which the false pendulum of a verge movement may be seen. Very seldom seen on everyday longcase clocks

Pillars The columns, often decoratively turned, which couple together the plates of a clock. In nearly all longcase movements they are permanently rivetted to the back-plate, but pinned or latched to the front plate

Pineapple A type of finial rarely found on longcase clocks

Pinion A cog on an arbor consisting of leaves as opposed to teeth, which engage with the teeth of a wheel. Other than in the motion work, it is the wheel which drives the pinion and not the other way round. Pinions are highly polished and have rounded ends to the leaves

Pin Pallets Pallets on a pin-pallet escapement that are D-shaped in cross-section

Pin Wheel The second wheel in the strike train. It has eight pins equidistantly round its rim, at right angles to the plane of the wheel; they trip the tail of the hammer during striking

Pipe An extended boss through which a hole is bored. A tube. A seconds hand in a longcase clock is mounted on a pipe which is split at its inner end so that the pipe can be pushed on to the tapered arbor of the escape wheel. The hour-wheel pipe is a much larger affair, described in the text

Pitch Circle A circle joining all points of contact between wheel-teeth and pinion leaves. That centred on the wheel would thus have a radius from the wheel centre to the point of contact only, not to the tip of the tooth

Pivot The trued and hardened end of an arbor which runs in a bearing

Pivot Hole The bearing of a pivot. When a worn pivot hole is repaired the process is known as re-bushing

Plates The two pieces of brass which support the arbors of clock trains. Some old clocks have split plates for ease in dismantling one train without disturbing the other

Play *See* Shake

Plinth The base section of a longcase clock. Sometimes fitted with skirting, sometimes standing on bun feet

Post or Stud A small stub axle mounted in the front plate to carry a wheel or a lever

Pulleys The insertion of a pulley on the line results in the downward travel of the weight being halved, and also the effective pull on the barrel being halved. Thus the pull of a 12lb weight, disregarding friction, would be 6lb, and a line run-out from the barrel of 6in lowers the weight by only 3in

Pump Action Where two bells are used in some forms of striking, the double hammer mechanism has to change from one to the other. This is often done by arranging pins on *both* sides of the pin wheel, and the pump action selects one or the other to operate one or other of the hammers. Pump action is often affected by a fence cam on the minute wheel or motion wheel (*See* Fence)

Push-on Fit *See* Friction fit

Quarter-hour Circle or Ring The innermost ring of early chapter rings, showing the quarter-hour divisions. Used mostly when clocks had one hand, before dials showed the minutes, but it took some time to die out. On some quarter-hour rings a second ring divided each quarter into three, ie five minutes, and it was then possible to tell the time to within a few minutes; but the mere thickness of the single hand covered some 2½ minutes on the dial

Rack A toothed, curved lever to regulate the number of blows made by a bell-hammer. Its tail falls on the snail – a cam which 'sets' the number of teeth to be gathered up by the gathering pallet. The system was invented in 1676 and is still universally used

Rack Hook A one-toothed hook which positions the rack for striking. Its arm is raised by the lifting-piece, disengaging the hook and so allowing the rack to fall, prior to striking

Rack Pin The pin on the rack which locks the gathering pallet

Rack Spring A light spring anchored on the front plate and pressing on a short projection on the rack; it ensures that the rack falls correctly to the full extent determined by the snail

Rack Striking As opposed to count-wheel or locking-plate striking

Rack Tail The lower arm of the rack, on which is mounted a shaped stud or pin which falls on the steps of the snail. It is flexible or has some device to clear the snail if necessary (*See* Section 82)

Ratchet A series of teeth, set round the rim of a ring or a barrel. Each tooth is nearly vertical on one side and sloping on the other. A ratchet is fitted with a pawl or click which fits the teeth so that the ratchet can move in only one direction

Ratchet Wheel Never found on ordinary longcase clocks. It is the toothed wheel mounted on the square of the arbor of the mainspring of a clock, and is used to set up the tension of the spring

Rating Nut Perhaps the most important nut in a pendulum clock, for it controls the length of the pendulum and so the rate at which the clock operates. On a longcase clock it is immediately below the bob of the pendulum. Screwing it upwards makes the clock go faster and vice versa

Re-bushing The process of fitting a new bush, bouchon or pivot bearing in the plate of a clock

Recoil The general form of escapement for a grandfather clock. The pallets, having caught the tooth of an escape wheel, continue to move towards the wheel, pushing it slightly back — hence the name recoil

Regulator An especially accurate clock, often a longcase, so-called because clock-makers and repairers regulated their clocks by one

Retaining Pin A pin, usually very slightly tapered, put through a hole in the end of a column or stud to hold the part in place or prevent any side play

Rise and Fall A method of raising the pendulum up or down for regulation. Not found on the ordinary grandfather clock

Round Broach *See* Broach

Roundel A disc or convex circular plate, usually engraved, mounted in the centre of the break-arch of a dial plate. The term is heraldic also, when roundels of metal or colours each have separate names. Roundels on longcase clocks are seldom plain, either bearing the words Tempus Fugit or some form of engraving encircling the maker's name

Seat-board The board, usually about an inch thick, carrying the movement of a longcase clock. The movement is bolted to the board, and the latter rests on the cheeks of the case. The seat-board is cut to clear the pendulum rod at the back, and is holed to allow the cables free play along the barrels

Seat Bolts The two bolts holding the clock movement to the seat-board, often called holding-down bolts

Seconds Dial A small dial or ring beneath the 12 on a clock dial; its hand indicates the seconds. A seconds ring

Seconds Hand A small hand, often balanced about its axis, mounted on a little friction-fit pipe which fits on the end of the escape-wheel arbor. It shows the seconds and will indicate the recoil in that type of escapement (*See also* Sweep seconds)

Serpentine Hands Hands shaped to be alternately convex and concave, ie as a serpent crawls. Found on longcase clocks from the eighteenth century onwards

Seven-and-a-half-minute-marks Small marks, perhaps like little arrowheads, to indicate the half of a quarter-hour. There are only four and they are rarely found on more modern clocks. Usually positioned just outside the minute ring, or within it if the Arabic minutes are within it

Shake A term used to describe the play or freedom of movement between two parts of a mechanism. A pivot, for instance, should fit its hole perfectly with no looseness or lateral shake. It can and should slide in and out of its bearing: it has limited end-shake but no side-shake

Silvering, Silvered By a process described in textbooks, the metal silver is deposited on brass or copper surfaces. A brass dial may have a silvered chapter ring and other rings or roundels. A silvered dial means the whole dial plate, and it would generally be engraved, chapters and all

Slide-up Hood Some seventeenth-century longcase hoods were raised up and fixed by a latch, in order to get at the clock dial. Now very rare

Snail A snail-shaped cam; an integral part of rack striking which limits the hammer blows to the number indicated by the hour hand. It is fixed to the hour wheel mounting

Spandrels The corner-pieces or decorations of the dial of a clock. They take many forms; cast-brass ones are usually gilded. Painted dials often depict the four seasons, or perhaps have geometrical designs. Brass gilded spandrels were to some extent governed by fashion and can help in dating a clock. Very fine clocks may have solid silver spandrels

Spires A type of finial found on longcase hoods from about 1750 onwards

Spoon A sort of bolt, usually of wood, for locking the hoods of longcase clocks

Star Wheel A twelve-point star-shaped disc on which a snail is mounted, on a repeating clock. Not found on most longcases. Also, on very old longcases and lantern clocks a star wheel is used to operate the lifting-piece, being mounted integrally with the hour wheel on the hour-wheel arbor (*See also* Jumper)

Steady-pins Where a bridge or a cock is screwed to a plate, any lateral or rotary movement is prevented by rivetting one or two steel pins or small lugs to the piece which fit into holes in the plate

Strap Hinges The type of hinge on the trunk door of a longcase clock. One short arm is screwed to the case and a long one, usually decoratively cut, to the inside of the door, angled round the door edge

Strike-silent A dial or other indicator which can be manually turned to eliminate the striking of the clock, at night for instance. The dial shows the two words and a hand may be pointed to whichever is desired. Some clocks – not often longcases – have a slot and lever

Strike Train The train which operates the strike, consisting normally of four wheels and a fly. There are two or three types of train. All are motionless except when let off by the going train to operate the bell-hammer

Stud *See* Post

Suspension Spring The small flat steel spring which suspends the pendulum from the back-cock or other mounting. A brass block at one end fits into the crutch, and the pendulum rod is screwed into it

Swan-neck The decoration at the top of the hood of a longcase clock. Two swan-neck pieces of carved wood face inwards towards the central finial

Sweep Seconds Some longcase clocks, especially those with a dead-beat escapement, are fitted with a centre seconds hand. It is long and balanced, reaching to the minutes circle. A badly balanced hand can prevent the clock going properly. The length should be just short of the minute hand, so that the latter may be moved without interfering with the sweep-seconds hand. The pipe of a sweep-seconds hand is best trued in a lathe to remove all surplus metal, because the hand should be as light as possible. Nowadays one can make serviceable centre-seconds hands out of duralumin, and enamel them black.

Brass ones tend to be heavy and springy which can result in an undesirable amount of whip. Where aluminium is used, the balance has to be achieved by filing carefully, starting with the thick end heavier than the thin or pointed end. In brass types, balance is readily effected by adding or removing solder to or from the back of the thick end. Both take much time and care

Third Wheel The wheel which engages the pinion on the escape-wheel arbor. In clocks of more than eight days' duration there are more wheels in the train, but the third wheel sometimes retains its name by common usage

30-Hour Clock A type of longcase clock usually wound up by pulling on a rope or chain. Some had false winding-holes in the dial as a status symbol! More rarely they were wound by key. Generally without a seconds dial

Timepiece A clock which tells the time only, without any striking or other attachment. It therefore has only one train (*See also* Trip strike)

Tooth The rims of wheels which engage with pinions are cut into small projections, called teeth. The equivalent on a cog or pinion is leaf or leaves. Teeth are of different shapes – a study in itself. Escape wheels have quite different teeth from those of any other wheel, and themselves have many differences. The study of wheel-teeth and pinion leaves is an essential part of engineering

Torque A twisting stress set up by pressure in an object such as a rod, a plate or a bar; or a force applied to a fixed object which tends to twist that object. The mechanics of the lever reveal that such a force may be transmitted or resolved in many directions in a machine such as a clock

Train A series of wheels and pinions engaging with one another for some purpose. A clock has a going train, a strike train, a chime train, an astronomical train, etc

Trip Strike Rarely found on longcase clocks, and then usually on 30-hour clocks. The movement is not that of a striking clock, but has a bell and its hammer. A pin on the minute wheel trips a lever, or the hammer tail, at each hour so that only one blow is struck on the bell. Also known as passing strike, common on skeleton clocks

Trunk The part of a longcase between the hood and the plinth or base; the door is mounted on it

Twist Pillars *See* Barley twist. Fine specimens have both right- and left-hand twists for the sake of symmetry

Veneer Thin sheets of fine-grained wood used over the surface of clock cases and other furniture. An oak case would be veneered with walnut or mahogany, or even oak of a different grain — to form, say, a herringbone effect. Where veneer is only partial and is let into the surface, it is termed inlay

Verge The word means 'staff', hence the man who carries the staff in church is known as the verger. In clocks it is the staff or arbor which carries the pallets of a crown-wheel escapement. The pallets are set at right angles on the staff, each pallet over opposite sides of the crown wheel. Found only on very old longcase clocks, generally with short bob pendulums. There are, however, French longcase clocks — Comtoise, Morbier, Morez, etc — which embody verge escapements, the older ones with the crown wheel upside-down and very long pendulums

Warning Wheel The fourth wheel in the strike train. It carries a pin which is caught just before each hour (or half-hour too on some clocks) by the lifting-piece. When this pin is caught the clock is said to warn that it is about to strike, which it does directly the lifting-piece falls

Weights On a longcase clock the motive force is the weight, and there are usually two, one for each train; that of the strike train is slightly the heavier to avoid unduly slow striking. The poundage varies according to the type of clock, from 8lb to 14lb for 8-day clocks. Chime trains need more, and month, 3-month or year clocks even more still, together of course with stronger cables

Wheel A toothed disc fixed to an arbor either with or without a pinion attached. The spokes of the wheel, if any, are called crossings

Winding Holes The holes in the dial plate that permit the key to be fitted on to the winding squares

Winding Squares The squared ends of the barrel arbors which protrude through the front plate to the level of the surface of the dial, or just short of it. Where there is bolt-and-shutter maintaining power, the ends of the squares are short of the dial to allow the shutters full clearance

Year Clock A clock which runs for a year on one winding. It goes twelve times longer than a month clock, ie 12×32 days = 384 days. Antique year clocks are very rare indeed and coveted museum-pieces

APPENDIX

REPLACING A BROKEN GLASS IN THE HOOD DOOR OF A LONGCASE CLOCK

Remove the door from the hood. Having fully and carefully detached all the old putty and bits of broken glass from the door, see that the frame is strong and rigid. Any loose veneers should be glued back under protected pressure.

In a square frame, replacing a glass presents no difficulty. It is quite hazardous, however, for an amateur to try to cut a glass for a break-arch frame. One cannot make a right-angled cut in a piece of 'picture' glass, which is the glass needed here. One way to solve the problem is to obtain a large sheet of thin white cardboard and make a template exactly the size of the glass needed to fit the frame. Allow about 2 or 3mm clearance all the way round.

The outer veneer or front face of the frame is cut to form the break-arch, with a sharp angle each side where the arc of the circle meets the straight edge of the top of the frame. The back of the inner frame, which is usually of oak, has no such sharp angle; it is cut away to form a pronounced curve. It is to this curve that the glass must conform. The template must therefore be cut to this curve, allowing 2 or 3mm clearance. Having fitted the template and cut it to shape, mark it on one side to show 'back' or 'front'. This is not necessary if the frame is well made, only if it is slightly out of true.

Now remove the template and lay it on the back of a sheet of hardboard firmly supported on a flat bench top. (Or you could use a firm flat table with a tablecloth stretched over it.) Strict cleanliness is essential.

Place the new sheet of glass over the template so that one corner and two sides of each coincide. This will save cutting a) one side, and b) the bottom edge of the new glass. Now with a good glass-cutter and a steel rule as guide — a builder's steel angle is suitable — cut the glass over the side of the template right across the pane, past the place where the break-arch will meet the side cut.

You now have a straight line score-cut across the pane of glass. Lift up the latter and with the end of the cutter-head gently tap the score-mark from *beneath* the cut. Tap-tap it all along the mark, at the same time watching for signs of 'crack' — it looks as if minute air cracks are occurring all along the cut. The tapping should be gentle and exactly beneath the crack. When it appears to be complete and ready to be severed, hold the pane of glass as if it were made of paper and you were about to tear it along a downwards line; hold the pane with the thumb and forefinger of each hand, *one hand on each side of the cut.* Now break the pane as if you were breaking a biscuit, or a flat wood batten, and you will find that the 'crack' will *run* cleanly down the whole length of the pane.

This may not sound right but it works very well, as you will discover by doing tests on odd bits of the old broken glass.

Alternatively, you can lay the pane on the sharp edge of the steel rule so that the 'cut' and the edge coincide, and with the part to be cut off not directly supported. A firm, sharp downward pressure on it should result in a clean break.

Replace the pane on the template in order to cut the break-arch shape. Hold it firmly so that neither it nor the template can move. With the glass-cutter press down firmly on the glass, with the wheel (or diamond) exactly over the edge of the template beneath the glass. Trace the line of the edge of the card right round to the opposite side of the pane. See that the scored cut goes fully to the edges each side. Raise up the glass and gently tap-tap it from beneath as before, but do *not* attempt yet to remove the surplus.

Lie the pane down again and very carefully cut extended radial lines at 40 or 50mm intervals from the *outside* of the curve to the edge of the pane of glass. Tap all of these to ensure the cut is right through the glass, but take care that the radial lines do not cross the arch curve. This is important.

When all cuts have been fully tapped, break off outer pieces. During this tapping a piece or two may already have parted. Clean up all sharp edges with emery paper.

Now lie the pane on the template and according to your mark 'back' or 'front', transfer the pane to the frame. It should now fit nicely with slight clearance all round. Remove the glass, support the frame beneath the veneered edge, and bed a thin strip of putty all round where the glass will lie. Press the glass in place and bed it down evenly all round, supporting the veneer whenever and wherever pressure is made. Any excess putty exuded may be removed later.

Now fix the glass in place with four to six small panel-pins driven into the frame as when framing a picture. Finish by pressing putty over the inside edges, like a window frame, in such a way that it cannot be seen from the outside. Remove any which may have exuded on the front of the glass, and trim it all neatly.

The putty may be stained with wood stain to match the frame as nearly as possible, and left to dry, preferably for several days, before the hood door is replaced.

When the putty is dry enough for the door to be handled, clean the glass with window cleaner, and clean up all the woodwork, veneering, etc, with a good furniture polish.

The job is now complete.

Did you have to go and get another pane of glass? We hope not.

REFERENCE SOURCES FOR THE WORKSHOP

Britten, F. J. *The Watch and Clockmaker's Handbook, Dictionary and Guide* Spon (1946) (Especially for anchor escapement, black waxing, depthing and silvering a dial)

Cescinsky, H. and Webster, M. R. *English Domestic Clocks* Hamlyn (1969) (Especially for hands)

Gordon, G. F. C. *Clockmaking Past and Present* Technical Press (1949) (Especially for anchor escapement, hands and depth tool)

Wilding, J. *How to Repair Antique Clocks* Brant Wright Associates (1979) (In two volumes, both excellent for studying detail, Vol 2 especially for bushes and re-bushing, etc)

'Hands and their Makers' *Clocks* Vol 3, no 1 (July 1980)

'Don't Beat About the Bush' *Clocks* Vol 4, no 11 (May 1982)

'Polish Makes Perfect' *Clocks* Vol 4, no 9 (March 1982)

'Horrors from an Horologist's Workshop' *Clocks* Vol 6, no 3 (December 1980)

'A Watchmaker's Workshop Regulator' *The Horological Journal* Vol 124, no 4 (October 1981) (Illustrating three types of depth tool)

'Making a Depthing Tool: Appendix 1' *Timecraft* Vol 2, no 4 (April 1982) (Showing a different type of depth tool from that described in this book)

RECOMMENDED BOOKS

General Study

Barker, D. *The Arthur Negus Guide to English Clocks* Hamlyn (1980)
Britten, F. J. *Old Clocks and Watches and their Makers* Eyre Methuen (1973)
Britten, F. W. *Horological Hints and Helps* Technical Press (1977)
Cescinsky, Herbert *Old English Clocks* Routledge (1938)
Cescinsky, H. and Webster, M. R. *English Domestic Clocks* Hamlyn (1969)
de Carle, Donald *Practical Clock Repairing* NAG Press (1977)
Edwardes, Ernest L. *The Grandfather Clock* John Sherratt & Son (1971)
Gordon, G. F. C. *Clockmaking Past and Present* Technical Press (1949)
Loomes, Brian *White Dial Clocks* David & Charles (1981)
Roberts, Deryck *The Bracket Clock* David & Charles (1982)
Symonds, R. W. *A Book of English Clocks* Penguin (1947)
Symonds, R. W. *Masterpieces of English Furniture and Clocks* Batsford (1940)
Wilding, John *How to Repair Antique Clocks* 2 Vols, Brant Wright Associates (1979)

Further Study

Britten, F. J. *The Watch and Clockmaker's Handbook, Dictionary and Guide* Spon (1946)
Bruton, Eric *The Longcase Clock* Arco Publications, New York (1970)
de Carle, Donald *Clock and Watch Repairing* Pitman (1969)
de Carle, Donald *The Watchmaker's Lathe* Hale (1971)
Fleet, Simon *Clocks Pleasures and Treasures* Weidenfeld & Nicolson (1961)
Gazely, W. J. *Clock and Watch Escapements* Heywood (1956)
Gazely, W. J. *Watch and Clockmaking and Repairing* Heywood (1953)
Jagger, Cedric *The World's Greatest Clocks* Hamlyn (1977)
Lloyd, H. Allen *The Collector's Dictionary of Clocks* A. S. Barnes (1964)
Lloyd, H. Allen *Old Clocks* Benn (1964)
Rawlins, A. A. *The Science of Clocks and Watches* Pitman (1969)
Smith, Alan *The Country Life International Dictionary of Clocks* Country Life (1979)

General Interest

Allix and Bonnert *Carriage Clocks* Antique Collectors' Club (1974)
Baillie, H. G. *Clockmakers of the World* Vol 1, NAG Press (1966)
Beeson, C. F. C. *English Church Clocks* AHS (1971)
Bruton, Eric *Clocks and Watches* A. Barker (1967)
Cumhail, P. W. *Investing in Clocks* Barrie & Rockliff (1967)
Daniels, George *The Art of Breguet* Sotheby Parke Bernet (1975)
Edey, Winthrop *French Clocks* Studio Vista (1967)
Edwardes, Ernest L. *Weight Driven Chamber Clocks* John Sherratt & Son (1971)
Fleet, Simon *Clocks* Octopus Books (1972)
Goudge, Elizabeth *The Dean's Watch* Hodder & Stoughton (1960) – Novel
Joy, Edward T. *The Country Life Book of Clocks* Country Life (1967)
Loomes, Brian *Clockmakers of the World* Vol 2, NAG Press (1976)
Quill, H. *The Man who found Longitude* John Baker Publishers (1966)

Robinson, T. R. *Modern Clocks, Their Design and Maintenance* NAG Press (1934)

Royer-Collard, R. B. *Skeleton Clocks* NAG Press (1969)

Tyler, E. J. *European Clocks* Ward Lock (1968)

Ullyet, Kenneth *Clocks and Watches* Hamlyn (1971)

Ullyet, Kenneth *In Quest of Clocks* Hamlyn (1968)

Antiquarian Horology The quarterly journal of the Antiquarian Horological Society

Clocks Monthly magazine, Model & Allied Publications Ltd

Horological Journal Monthly official journal of the British Horological Institute

Timecraft Monthly magazine, Brant Wright Associates Ltd

ACKNOWLEDGEMENTS

Thanks are due to my wife for her support and patience, to my partner Gordon Craig for his help, to William Payne for the fine photographs, and to A. Blackbourn for the excellent drawings. Finally my thanks are due in no small measure to my publishers for their kind co-operation and help.

INDEX

The figures refer not to page numbers but to the book's numbered sections. See also pages 9–11 for a description of the parts of a longcase clock.